Environmental Ergonomics

Environmental Ergonomics: Principles, Methods, and Applications provides the philosophy, principles and application of environmental ergonomics as a universal concept and considers total environments as an integration of environmental factors to which people are exposed.

The book develops the definition of environmental ergonomics and presents the principles, methods and application of knowledge in the areas of human response to heat and cold, sound, vibration, light and air quality and addresses diverse environments and people. The title explains the effects of the environment on the health, comfort and performance of people for all environmental components, introduces environmental ergonomics as a universal subject and offers general principles and methods for measuring and representing the environment and its effects. A wide range of case studies demonstrates the application of the subject for all environmental components and for integrated environments. Special environments such as vehicles and unique populations such as people with disabilities, are considered in the book. The title concludes with an in-depth understanding of total environments with a description of how to conduct an environmental ergonomics survey and a case study.

An ideal read for any professional or student interested in environmental ergonomics, including architects, occupational hygienists, interior designers, systems engineers, civil engineers, HVAC engineers and building services engineers. The title will help any reader develop a thorough understanding of the effects of the environment on the health, comfort and performance of people.

Environmental and Occupational Health Series

Series Editors

LeeAnn Racz
Bioenvironmental Engineer, US Air Force, Brandon, Suffolk, UK

Adedeji B. Badiru
Professor and Dean, Graduate School of Engineering and Management, AFIT, Ohio

Handbook of Respiratory Protection
Safeguarding Against Current and Emerging Hazards
LeeAnn Racz, Robert M. Eninger, and Dirk Yamamoto

Perfluoroalkyl Substances in the Environment
Theory, Practice, and Innovation
David M. Kempisty, Yun Xing, and LeeAnn Racz

Total Health Exposure
An Introduction
Kirk A. Phillips, Dirk P. Yamamoto, and LeeAnn Racz

Forever Chemicals
Environmental, Economic, and Social Equity Concerns with PFAS in the Environment
David M. Kempisty and LeeAnn Racz

Environmental Ergonomics: Principles, Methods and Applications
Ken Parsons

More Books Forthcoming

More about this series can be found at: https://www.crcpress.com/Environmental-and-Occupational-Health-Series/book-series/CRCENVOCCHEASER

Environmental Ergonomics

Principles, Methods, and Applications

Ken Parsons

CRC Press
Taylor & Francis Group
Boca Raton London New York

CRC Press is an imprint of the
Taylor & Francis Group, an **informa** business

First edition published 2025
by CRC Press
2385 NW Executive Center Drive, Suite 320, Boca Raton FL 33431

and by CRC Press
4 Park Square, Milton Park, Abingdon, Oxon, OX14 4RN

CRC Press is an imprint of Taylor & Francis Group, LLC

ISBN: 978-1-032-50332-5 (hbk)
ISBN: 978-1-032-51387-4 (pbk)
ISBN: 978-1-003-40196-4 (ebk)

DOI: 10.1201/9781003401964

Typeset in Times
by SPi Technologies India Pvt Ltd (Straive)

Contents

Part II Thermal Environments

Part III Environmental Ergonomics and the Assessment of Noise and the Acoustic Environment

Part V Light and the Visual Environment

Part VI Environmental Ergonomics and Air Quality

Part VIII Environmental Ergonomics and Integrated Environments

Preface

Environmental Ergonomics is the application of knowledge of human response to the environment to the design of systems. All people are surrounded by environments, at all times, usually on the ground breathing air, indoors or outdoors and wearing clothing, but also in more diverse environments from under and on the water, up mountains, down holes, in the sky, in buildings, vehicles, protective clothing microclimates, in space and eventually on other planets and more. Where people are involved there will be a human system and this will include a human environment.

The aim of this book is to provide the philosophy, principles and application of environmental ergonomics as a universal concept and approach to application. It considers total environments as an integration of environmental components to which people are exposed. The book develops the definition of environmental ergonomics and presents the principles, methods and application of knowledge in the areas of human response to heat and cold, sound, vibration, light and air quality as well as a consideration of more diverse environments and people. It considers the effects of the environment on the health, comfort and performance of people for all environmental components.

Part 1 presents the introduction to environmental ergonomics as a universal subject within the context of Ergonomics and presents general principles and methods for measuring and representing the environment and its effects focussed on application. Parts 2–6 consider thermal environments; sound (noise); whole-body and local vibration; vision and lighting; and air quality. Part 7 considers diverse environments and populations. It includes air pressure, extra-terrestrial environments and transient environments, vehicles and more as well as people from around the world and people with special requirements such as those with physical disabilities.

Part 8 integrates the knowledge provided in the book to consider total environments with a description of how to conduct an environmental survey and a case study.

Case studies are used to demonstrate the practical application of environmental ergonomics. Although hypothetical, they are based upon experience of actual environmental ergonomics assessments. Hypothetical names and characters are included to provide context and realism and do not represent actual people.

Ken Parsons
Loughborough, 2024

Acknowledgements

To
Jane, Benjamin, Samuel, Hannah, Anna, Mina, Richard,
Nancy, Thomas, Edward, Mary.
Mother; Florence, Father; Eric, Sister; Christine, Brother, Ean

About the Author

Ken Parsons is Emeritus Professor of Environmental Ergonomics at Loughborough University in the United Kingdom and has spent over 40 years in research and its application concerned with human response to the environment. He joined Loughborough University in October 1971 as an undergraduate, having specialised in mathematics, physics and biology at school. He read 'Ergonomics' in the internationally respected and innovative department of Ergonomics. The then Department of Ergonomics and Cybernetics was established in 1959 by Professor Will Floyd who brought together experts from around the world to cover the requirements of each of the subject components of Ergonomics. Environmental Ergonomics was taught by experts in noise and vibration, vision and lighting and the thermal environment all with facilities to provide practical experience. The department developed into the Department of Human Sciences to recognise the significant contributions in human biology and psychology as well as Ergonomics and finally Ergonomics moved to the School of Design (later Design and Creative Arts) at Loughborough. Ken graduated in Ergonomics in 1974 having completed a project using multi-dimensional scaling subjective techniques to determine physical factors that influenced the quality of lighting in lecture theatres, supervised by Susan Harker and Peter Stone.

After graduating Ken moved to Hughes Hall, Cambridge University, to gain, with a distinction, a postgraduate qualification in teaching (mathematics) and study education, much influenced by the work of Wittgenstein and Plato. He then moved to the Institute of Sound and Vibration Research at the University of Southampton, UK where he took Masters classes and passed examinations in Sound and vibration and conducted laboratory and field studies into human response to vibration to successfully complete his PhD. As a postdoctoral fellow, he carried out studies into the perception of vibration, with particular application to the responses of people to vibration in buildings. At Southampton Ken joined fellow Loughborough Ergonomics Alumni, Chris Lewis and Eleri Whitham all supervised by Mike Griffin head of the Human Factors Research Unit.

Ken returned to the Department of Human Sciences at Loughborough University in 1981, as a lecturer in Environmental Ergonomics, with specific responsibility for research and teaching in human response to thermal environments, building upon the work of the recently retired, Mike Cooke. He founded the Human Thermal Environments Laboratory at Loughborough in 1981 and was promoted to senior lecturer, reader and eventually professor in 1996. He gained a certificate in management from the Open University in 1993. Ken became head of the Department of Human Sciences in 1996 covering research and teaching in ergonomics, psychology and human biology. He was Dean of Science from 2003 to 2009 and pro-vice chancellor for research from 2009 to 2012. He was chair of the United Kingdom Deans of Science from 2008 to 2010.

In 1992, he received the Ralph G. Nevins award from the American Society of Heating, Refrigerating and Air-Conditioning Engineers (ASHRAE) for 'significant accomplishments in the study of bioenvironmental engineering and its impact on human comfort and health'. The Human Thermal Environments laboratory was awarded the President's Medal of the Ergonomics Society in 2001. He is one of the co-authors of the British Occupational Hygiene Society publication on thermal environments and has contributed to the Chartered Institute of Building Services Engineers publications on thermal comfort as well as to the *ASHRAE Handbook: Fundamentals*.

He has been a fellow of the Institute of Ergonomics and Human Factors, the International Ergonomics Association and the Royal Society of Medicine. He was a registered European Ergonomist and an elected member to the council of the Ergonomics Society. He has been a scientific advisor to the Defence Evaluation Research Agency and the Defence Clothing and Textile Agency and a member of the Defence Scientific Advisory Committee. He has been both secretary and chair of the thermal factors committee of the International Commission on Occupational Health (ICOH), chair of the CNRS advisory committee to the Laboratoire de Physiologie et Psychologie Environmentales in Strasbourg, France, and is a life member of the Indian Ergonomics Society. He was a visiting professor to Chalmers University in Sweden and has been a member of the committee of the International Conference on Environmental Ergonomics from its early days. He was an advisor to the World Health Organization on heatwaves and a visiting professor to Chongqing University in China, where he was a leading academic to the National Centre for International Research of Low Carbon and Green Buildings. He was scientific editor and co-editor in chief of the journal *Applied Ergonomics for 33 years* and is on the editorial boards of the journals *Industrial Health, Annals of Occupational Hygiene* and *Physiological Anthropology*.

He is co-founder of the United Kingdom Indoor Environments Group and a founding member of the UK Clothing Science Group, the European Society for Protective Clothing, the Network for Comfort and Energy Use in Buildings and the Thermal Factors Scientific Committee of the International Commission on Occupational Health (ICOH). His contribution to the development and production of Ergonomics standards spans over 40 years. He was chair of ISO TC 159 SC5 'Ergonomics of the Physical Environment' for over 20 years and is convenor to the ISO working group on integrated environments. He was chair of the British Standards Institution committee on the ergonomics of the physical environment from 1989 to 2024 and was convenor of CEN TC 122 WG11, which is the European standards committee concerned with the ergonomics of the physical environment, for over 30 years.

He has presented papers to over 200 national and international conferences and over 20 invited keynote addresses worldwide over five continents as well as acting as PhD examiner, and external assessor for professorial promotions in the UK, Asia, and across Europe. His over 200 published works include author of the seminal paper 'Environmental Ergonomics - A Review of Principles and Methods', 2000; Vibration discomfort (parts 1 to 4) with Griffin and Whitham, 1982; Vibration perception thresholds with Griffin; 1988 and The effects of gender, acclimation state, the opportunity to adjust clothing and physical disability on requirements for thermal comfort, 2002. He has made a significant contribution to all editions of the Environmental Ergonomics section of the book 'Evaluation of Human Work' (Wilson and Sharples) fourth edition 2015 and is the author of the books *Human Thermal Environments* (3rd Edition 2014); *Human Heat Stress* (2019), *Human Thermal Comfort* (2020) and *Human Cold Stress* (2021).

He is responsible for new terms and methods including 'the thermal audit', the subject 'Human Psychrometrics', the RDC – Regulations Distraction Capacity – method for considering the effects of the environment on human performance and environmental indices that represent the effects of environmental variables that influence human response as a single number that can be used in application. He has also pioneered psychophysiological homeostasis and the general application of the adaptive approach in considerations of human response to environments that includes the contribution of human behaviour to responses to the full range of environmental components as well as to total and integrated environments.

Part I

Environmental Ergonomics

Part 1 includes three chapters. It provides the definition, philosophy, principles and methods of environmental ergonomics (Chapter 1). It includes the measurement of the environment (Chapter 2) and the human response to it (Chapter 3).

Environmental ergonomics will be a new subject to many as expertise and knowledge into the human response to the environment has traditionally been compartmentalised into specific areas such as human response to noise or the control of lighting, conditions for thermal comfort and so on. Different cohorts have been involved in research and application, and a whole infrastructure of people, institutions and organisations has developed around each single component of the environment. It is one of the unique features of ergonomics that it is human-centred and that the environment is considered as a whole.

It is the human disposition (along with all other living things and objects) that they are continuously affected by, and respond to, the environment that surrounds them. Environmental ergonomics considers the whole and not just the sum of the parts or any individual component. Its development as a subject and in application, will greatly enhance the environments in which we live, work and play.

We should note that at the time of writing this book, we are at the dawn of a new era. People will occupy environments in the future that are not on Earth. There are numerous planned missions to space stations, to inhabit Earth's moon and beyond. Understanding the human response to those new environments and how principles can be applied is a challenge for environmental ergonomics in the future.

Part 1 provides the background and common principles and knowledge, methods for measuring the environment and methods for measuring the human response to the environment, which apply to the assessment of all environmental components.

DOI: 10.1201/9781003401964-1

It presents how they can be used to assess the effects of the 'total' environment on the health, comfort and performance of people.

Chapter 1 defines environmental ergonomics as the application of knowledge of human response to the environment to the design of systems (Figure P1). Emphasis is given to the whole system such that components are not viewed in isolation. Environmental components of thermal; sound; vibration; light and air quality, units of measurement and their integrated effects are described. Individual components of the environment have been systematically studied and researched for over 100 years; however, there has been only limited investigation into their integrated effects.

The French philosopher, Rene Descartes, considered the question of how we know that the external environment exists at all, given that all of the evidence we have for its existence is based upon often unreliable signals from our senses. He concluded that if we are asking the question then it must exist. Environmental ergonomics makes that assumption. We ride on the dimension of time observing the three spatial dimensions of up and down; side-to-side; and fore-and-aft (with roll, pitch and yaw providing the six degrees of freedom of motion); and to observe the four dimensions, we must have a fifth related to thought. Energy operates within that framework.

Claude Bernard was a French scientist who described how the body maintains an internal environment despite variations in the external environment in order to survive. Parsons (2020) and others have shown how this can be extended not only beyond a drive for physiological survival but also to a psychological response where comfort can be considered to be a controlled variable.

We can draw on the work of the Italian Scientist, Sanctorius, to link human response to the environment with the laws of physics. From his experiments and the later laws of thermodynamics, we can conclude that energy into and out of the body can be accounted for. There is nothing special about people. So, for a constant internal environment, the energy in the body must balance with energy from it. Hence, an audit can be made of energy quantities and pathways, a fundamental method of analysis and prediction of human response to the environment and often the basis for computer models.

Psychophysical laws and methods developed in the nineteenth-century relate levels of the physical environment to sensation and also provide a basis for predicting human response to the environment. Environmental ergonomics methods, including subjective; objective; behavioural and adaptive; and modelling methods are identified.

Chapter 2 begins with a discussion of energy and how it forms the basis for units used in environmental ergonomics to represent the physical environment to which people are exposed. Heat, mechanical, chemical and electromagnetic forms of energy are considered and the electromagnetic radiation spectrum is presented. Equivalent sensation curves showing the relative effects of environmental frequency or wavelength on people are presented both from experimental data and in idealised form, used in international standards and units, for comparison.

Properties of instruments for measuring the physical environment are presented and validity; accuracy; range; sensitivity; reliability; signal-to-noise ratio; specificity; response time and calibration are defined and discussed. Measurement of human behaviour includes adaptive methods and measurement of the social environment include quantitative; qualitative and discourse methods. A final note on ethical considerations is provided.

Chapter 3 presents a simple model where a person continuously detects the physical environment, interprets signals from the sensors of the body and responds to them either subconsciously or consciously. Most people have the ability to report opinions and how they feel and subjective methods are described along with subjective scales and nominal, ordinal, interval and ratio levels of measurement.

Objective methods involve measurements of the response of the body and heart rate; body temperature; sweat rate; pupil diameter; body acceleration; urine and blood analysis and more are identified as well as performance at tasks. Behavioural methods are described particularly in terms of adaptive opportunities; and modelling methods to predict human response, often involving computer simulations, are discussed. The possibility of an integrated environmental index, to indicate environmental quality, for example, is discussed. It is emphasised that methods will be used in combination and will be selected and adapted to the context of any environmental ergonomics investigation. Figure P1 shows how environmental ergonomics contributes to ergonomics and a systems approach.

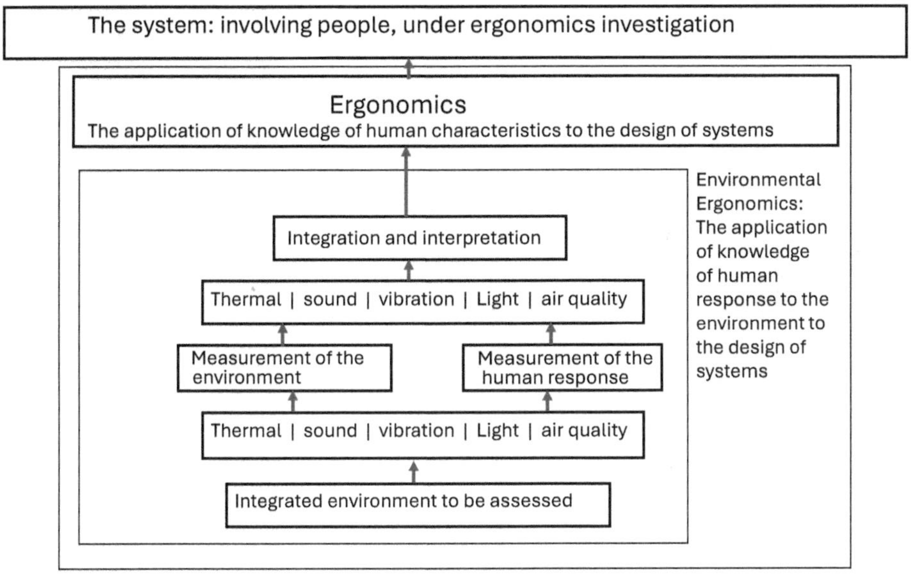

FIGURE P1 Environmental Ergonomics and its role in Ergonomics application.

1 Environmental Ergonomics

Environmental ergonomics is the application of knowledge of human response to the environment to the design of systems.

People live in a continuously changing environment around their body and continuously respond to it so that they can survive; remain healthy; gain comfort; satisfaction, inspiration and pleasure; and perform tasks effectively. Different systems have different objectives and environmental ergonomics uses knowledge of human response to the environment to contribute to the achievement of those objectives. It is part of design that also includes evaluation and assessment. Much of environmental ergonomics application is concerned with evaluating and assessing existing environments.

The environment exists as a totality but it can be conveniently represented as an integration of components such as sound, vibration, heat and cold, light, air quality and more. Much of the environment is detected by the skin (heat and cold, touch, pressure); it is also detected by specialist organs in the body (ears, eyes, etc.). Some environmental components appear not to be detected at all or have no, or unknown, effects, for example, magnetic fields, radio waves, cosmic radiation, or those at energy levels below the absolute thresholds of the sensors of the body.

Wilson and Sharples (2015a) and Parsons (2000) are among the few texts that have considered environmental ergonomics as an integrated subject. Classic texts that consider human response to environmental components include Parsons (2014) for the thermal environment as well as Fanger (1970) for his seminal work on thermal comfort; Kryter (1985, 1994) for noise and hearing; Griffin (1990) for his definitive work on human vibration (see also Mansfield, 2004); Boyce (2014) for the human factors of vision and lighting; and Fanger (1988) for consideration of air quality in terms of human perception.

Environmental ergonomics in its application is supported by numerous national (e.g. ANSI; DIN; BSI), regional (e.g. EN for Europe) and international (e.g. ISO) standards as well as those of professional institutions (e.g. ASHRAE, CIE). ISO 28802 (2012) was the first standard to consider environmental ergonomics as an integrated subject and presents basic practical information to allow an environmental ergonomics survey to be carried out.

Some environmental components, not always of primary concern but rise in prominence from time to time, receive popularity with apparent evidence of effects. Infrasound was said to travel long distances and become perceptible and annoying to some people with a disturbance to health. Electrical cables were blamed for infrasound that disturbed people.

Negative ions in the air were associated with pleasant and stimulating environments. (Sulaman, 1974). Their presence was found next to waterfalls, for example.

DOI: 10.1201/9781003401964-2

Positive ions were blamed for the increase in suicides and depression as they are carried by infamous winds such as the mistral in the south of France, the Foehn in Switzerland or the Chinook in California. However, for none of these effects has conclusive evidence been found for cause and effect (McIntyre, 1980).

In principle, it is the total environment that is the consideration of environmental ergonomics; however, in practice, environmental ergonomics is usually concerned with environments that elicit a measurable human response.

Environmental ergonomics is based upon a simple model that assumes that people detect their environment through senses, interpret signals from those sensors and provide actions (based upon the nature of the signals and influenced by experience) to preserve or achieve a desired state.

Detection of the environment by the human body can be conscious or unconscious and responses can be *automatic* (autonomic) such as sweating when too hot; *consciously determined* such as turning away from bright lights and glare; or part of the environment can be *imperceptible* with no apparent bodily effect but with possible consequences such as changes in mood or other affects not yet understood. People can also think about scenarios individually or discuss environmental conditions as a focus of social interaction and collective action.

ERGONOMICS

Environmental ergonomics is a part of ergonomics that can be defined as the application of knowledge of human characteristics to the design of systems (Parsons, 2015b). How people respond to environments is a human characteristic as are their size and shape, strength, agility, psychological capacity and disposition, anatomy and physiology, social interaction and more.

Knowledge of human characteristics is used to design systems so that they achieve their goals. Systems are defined by their limits and involve an interaction of components to achieve goals. Products and environments can be systems in themselves but are often examples of system components; therefore, a chair, protective clothing or a thermal environment are system components and the success of their design will be determined in the context of the goals of the overall system.

An aircraft seat has different requirements from an office chair. An office chair would not meet the safety requirements of an aircraft and the aircraft seat would not be suitable for day-to-day computer work in an office. Vibration limits in a helicopter may be mainly related to pilot performance whereas in a building they may be related to disturbance, anxiety and absolute perception. All should be considered in the context of the goals of the overall human system. Wilson and Sharples (2015a) note that methods used in ergonomics application should be selected, integrated and created (with innovation and inspiration) in the context of the system being designed or investigated.

It is useful to note that some definitions of ergonomics confound principles with ethical considerations. In short, it is perfectly possible to create an ergonomically designed torture chamber or extermination camp. It is, however, unethical to do so. Adding phrases such as '...to the design of systems to the benefit of people', etc. is laudable and ethical and makes an important point but it is not part of the definition of ergonomics or environmental ergonomics.

DETECTION AND PERCEPTION OF THE ENVIRONMENT

The external environment presents energy to the body in different forms representing environmental components. The detection and perception of specific environmental components will be presented in subsequent chapters. General and common factors relevant to all environmental components are presented in the following text.

The body can detect the absolute level of energy as well as the time-varying nature of energy level. The lowest level of energy the body can detect is called the absolute threshold. Below that level (sub-liminal), the environment exists but is not detected so it is reasonable to assume no conscious response.

As the energy level increases, a person will notice a change in level, and the threshold level for the change is called the just noticeable difference (JND). The JND may vary with level (and rate of change of level), and for environmental components made up of oscillating energy levels, such as sound and vibration, thresholds will also depend upon frequency (number of oscillations per second) as well as wavelength (interval or distance before repetition).

It is important to recognise that 'signals' from the sensors of the body that detect the environment and their interpretation are not directly related to the physical properties of the environment. Doubling the physical energy, for example, does not necessarily double the perceived intensity of the environment. Environmental ergonomics research determines the relationship between the physical environment and the human perception of that environment so that methods and units can be determined that can be used in the prediction of human response.

RENE DESCARTES (1596–1650) AND THE EXTERNAL ENVIRONMENT

Rene Descartes noted that evidence that we exist in an external environment cannot be derived from the evidence of our senses as the senses were not reliable and could easily be deceived. Hence, his famous phrase 'Cogito, ergo sum' ('I think, therefore I am'). The reason we know that we exist is the rational deduction that it is fundamentally true that we have asked the question of whether the external environment exists or not.

Environmental ergonomics assumes that a person has a mind (assumed to be the brain) that receives information from the body. The information received is not directly related to the external environment but there is some reliability and by both empirical and rational methods we can determine the relationship between the two and predict the likely human response of a person when exposed to a physical environment.

CLAUDE BERNARD (1813–1878) AND HOMEOSTASIS

Claude Bernard and others provide a reason why a person actively responds to external environmental conditions rather than simply accepting physical affects on the body directly related to the laws of physics. He provides an objective for human response and that is to preserve homeostasis. What he terms the internal milieu (the internal environment in the body) is held constant (is controlled) to allow the vital organs of the body to function in optimum conditions.

The internal environment of the body is not constant, however, but varies within controlled limits and if the external environment threatens the 'desired' controlled conditions at the time then the body responds to defend its desired position. This is consistent with the proposal of Claude Bernard who did not actually use the term homeostasis.

If the body becomes too hot and rises above a desired temperature of 37 °C, for example, then sweating will eventually occur to promote heat loss and restore homeostasis. Homeostasis is an example of autonomic control but pain, discomfort and more can also elicit a human response.

Parsons (2020) presents this autonomic and conscious or adaptive duality as a psychophysiological system of regulation and hypothesises human discomfort as one of the controlling factors that elicit human response in combination with homeostasis. That is there appears to be both physiological homeostasis and psychological homeostasis, one providing pleasure as a reward for achieving comfort and survival and the other also pursuing comfort and pleasure as a desired state.

SANTORIO SANCTORIUS (1561–1636) AND THE LAWS OF PHYSICS

Santorio Sanctorius provides an example of an important principle in environmental ergonomics and that is that people are subject to the laws of physics. He sat in a suspended chair for a long period and an attempt was made to account for all inputs and outputs to the body, in particular weight inputs and losses. The outcome was as accounted for. Nothing was created nor destroyed. The conclusion was that there is nothing special about people. They obey the laws of physics. This meant that a rational method for considering the effects of the environment on people could be used with the assumption that all energy could be accounted for. If homeostasis is to be maintained, for example, then all energy into the body must be balanced by all energy leaving it.

PSYCHOPHYSICS

Part of environmental ergonomics research uses psychophysics to underpin theory and methods. Guilford (1954) provides a history. He credits Herbart (1776–1841) with the concept of absolute threshold and Weber (1795–1878) with JND and Weber's law 'the just noticeable difference in a stimulus is proportional to the stimulus'. Fechner (1860) defined psychophysics as 'an exact science of the functional relationship between body and mind' and proposed a logarithmic relationship now termed the Weber-Fechner law. Stevens (1957) proposed a power law where the relationship between stimulus and sensation is an exponential function, where the exponent depends upon the type of stimulus. To determine the value of exponents, he developed the method of magnitude estimation where a participant would be presented with a stimulus and asked to provide a number (between 1 and 100 for example) that indicated the strength of the stimulus. Over a wide range of stimuli, an exponential relationship between stimulus strength and numerical estimation would provide the value of the exponent.

The use of psychophysics in environmental ergonomics can be considered as over-developed in concept (to a level not required for practical application) and too simplistic (direct models of the relationship between an environmental component and sensation are only a part of a complex relationship) for practical application in a total environment. For a fuller discussion of psychophysics, see Stevens (1975).

ENVIRONMENTAL ERGONOMICS METHODS

Knowledge of human response to the environment has been generated systematically for over 100 years to build an understanding to allow prediction of human response. Environmental ergonomics is concerned with the application of that knowledge to design appropriate environments or assess existing environments.

Empirical methods (strictly knowledge from experience) have exposed people to a range of environmental conditions and recorded human responses so that when encountered in the future a prediction of response can be made using equations often based upon averaged data. Data-based models (Parsons and Bishop, 1991; Parker and Parsons, 1990) have increased the utility of empirical methods by using matching techniques involving actual responses of people. If the final selection of human response by users of the model, involving the matching procedure, were recorded by the software, such that the model learns; then artificial intelligence can be included in the model.

Rational models are based upon an analysis using the laws of physics to simulate and predict human responses to the environment. Some methods combine rational and empirical approaches. Developments in mathematics and computing have allowed rational models to simulate the human in an environment leading to computer-aided environmental design. Physical models can simulate the human body and they range from simple spheres and heated manikins to simulated ears, eyes, nose and more.

In design, involving environmental development, evaluation or assessment, methods are required to quantify the environmental conditions and measure the human responses. In statistical analysis, the measure of the environment is called the independent variable and the measure of the human response is called the dependent variable.

Wilson and Sharples (2015a) provide a comprehensive range of ergonomics methods, many of which are appropriate for environmental ergonomics. They provide an edited work of 38 chapters covering many areas of methods used in ergonomics, including experimental design, methods for quantifying participant response as well as qualitative methods, human modelling, systems analysis, inclusive design, physiological measurement, sociotechnical systems design, participatory ergonomics and more. They point to how ergonomics has developed from the consideration of individuals in systems to the interaction of people with other people and artefacts in an environmental context. Environmental ergonomics has yet to take advantage of this development but has embraced adaptive approaches where people interact with environments rather than simply respond to them. This provides the requirement to adopt a broader range of methods and models than previously used.

Consideration of methods on the design and assessment of the physical workplace is provided by Wilson and Sharples (2015a). After an introduction by Sharples, Parsons (chapters 23 and 24) provides consideration of the environmental ergonomics survey and the assessment of the thermal environment. Howarth (chapter 25) follows with a chapter on the assessment of the visual environment and Haslegrave (chapter 26) considers the assessment of noise and the auditory environment. Interestingly a chapter by Feathers et al. (chapter 27) is included on anthropometry and ergonomics design and hence notes the relevance of the spatial environment in environmental assessment. Although this clearly influences the experience of a space, it has not traditionally been included in an environmental ergonomics survey.

THE POST-OCCUPANCY QUESTIONNAIRE

Completed buildings in use can be judged in terms of how much they meet their design specification and also how they satisfy user requirements and satisfaction. A process related to the environmental ergonomics survey but specifically concerned with building performance is the post-occupancy questionnaire which is part of the overall building appraisal. In principle, the post-occupancy questionnaire and assessment apply to all environments occupied by people and not only buildings.

Leaman (2003) notes that post-occupancy evaluation was first suggested in the 1970s in the USA to describe the process of assessing buildings in use from the occupants' point of view. A building can be assessed in terms of what it is intended to do (ends) rather than prescribing and assessing how it has been constructed (means). Environmental ergonomics considers how people respond to the environment (e.g. created by a building) in terms of their health, comfort and performance and although a post-occupancy assessment, carried out to judge the building performance after (maybe a year or more) occupation provides user experience to the total building and its functions there is clearly overlap with this and an environmental ergonomics survey when carried out in the building.

Post-occupancy assessments are designed specifically for a building in use and are often concerned with user satisfaction. Occupant performance, however, may also be considered. Bordas (2002) considers post-occupancy evaluation and has used self-assessment of performance by the occupants as a method for determining how people consider they are performing tasks.

A typical post-occupancy questionnaire is provided online by the University of Strathclyde (2010), Scotland. Reassurance of confidentiality and the ability to withdraw leads to participant agreement and general questions regarding gender, age and job. A scale is provided for all questions where a dot is placed in a circle to indicate a response from 1: very unsatisfactory to 5: very satisfactory (or a variation depending upon context). Each question is followed by a box providing an opportunity to give comments.

Questions on the building overall include building layout, meeting your needs, privacy, safety, ratings of specific areas, desk space, desk furniture, storage, interior finishes, effectiveness of shading (blinds), and toilets. Questions on the environment include summer indoor air temperature; winter indoor air temperature; ventilation and air quality; noise; natural daylight; artificial light; degree of environmental control; energy and water efficiency and whether they think it is important; cleanliness; overall quality; productivity; pleasure; inside to outside view; architectural design and landscaping; and friendly physical environment and how important this is. A box for additional comments is also provided.

Chapter 2 considers how to measure the environment that is experienced by people and Chapter 3 how to measure their response to it.

FURTHER READING

Boyce, P. R., 2003, *Human factors in lighting* (2nd Ed), CRC Press, Taylor & Francis Group, London and New York, ISBN 0-7484-0950-5

Fanger, P. O., 1970, *Thermal Comfort*; Danish Technical Press, Copenhagen, ISBN 87-571-0341-0

Fanger, P. O., 1988, Introduction of the olf and decipol units to quantify air pollution perceived by humans indoors and outdoors, *Energy and Buildings*, vol 12, pp 1–6.

Griffin, M. J., 1990, *Handbook of human vibration*; Academic Press, ISBN 0-12-303040-4

Kryter, K. D., 1985, *The effects of noise on man*, (2nd Ed), Academic Press, London, UK.

Mansfield, N. J., 2004, *Human response to vibration*, CRC press, www.crcpress.com, ISBN 0-415-28239-X

Parsons, K. C. 2014, *Human thermal environments* (3rd Ed), CRC Press, Taylor & Francis Group, Boca Ratan, Oxford, New York ISBN 978-1-4665-9599-6

Parsons, K. C., 2000, Environmental ergonomics: A review of principles, methods and models, *Applied Ergonomics*, vol 31, 6, pp 581–594

Wilson, J. R. and Sharples S., 2015, *Evaluation of human work* (4th Ed), CRC Press, Taylor & Francis Group, Boca Ratan, New York, Oxford, ISBN 978-1-4665-5961-5

STANDARDS

ISO 28002, 2012, Ergonomics of the physical environment -The assessment of environments by means of an environmental survey involving physical measurements of the environment and subjective responses of people. ISO, Geneva.

2 Measuring the Environment

THE PHYSICAL ENVIRONMENT

The physical environment that surrounds people and is of interest to environmental ergonomics is made up of energy in the form of heat, mechanical energy and electromagnetic radiation. An additional avenue relevant to air quality is the chemical interaction in the nose and mouth and sometimes on the skin. A more profound assessment would delve further into the laws of physics but for most practical applications, those will suffice.

The laws of thermodynamics show that for all forms of energy, we can use a common system of units based upon the Joule (J). Energy can be considered to be the capacity to do work. Work is done when energy is transferred from one place to another. One Joule is equal to the amount of work done when a force of 1 Newton displaces a mass of 1 kg through a distance of 1 metre in the direction of the force applied. It is also the energy dissipated when an electric current of one ampere passes through a resistance of 1 ohm for 1 second.

One calorie is equal to 4.184 Joules and is the heat to raise 1 gram of water by 1 °C. The Watt is a measure of power where 1 W = 1 J s^{-1}. One Joule is also equivalent to 1 kg m^2 s^{-2}; 1 N m; 1 Pa m^3; 1 W s and 1 C.V (Coulomb.Volt or Watts = amps × volts).

One erg is the work done by 1 dyne over a displacement of 1 cm (i.e. energy in the, no longer used, cm:gram:sec (CGS) system). I Joule = 10^7 erg. The term energy (en-erg-y) is derived from the erg and ergonomics (erg-o-nomics) was derived from the natural laws of work (it was invented/adopted as a name for a professional body in 1949 (Murrell, 1973), but clearly the study of people at work has been conducted for thousands of years).

In environmental ergonomics (mechanical), vibration is usually represented and measured as acceleration (m s^{-2}); sound as pressure (Pa) and thermal environments and electromagnetic radiation (e.g. infra-red and light) directly as Joules or as a rate of heat transfer in Watts.

Due to historical development and convenience of application, the physical units used to quantify a physical environment are often adjusted to accommodate the human response. Light, for example, is a part of the electromagnetic spectrum (Figure 2.1) inherently linked with vision and the human eye, and hence units include the spectral properties and sensitivity of the eye and visual system (e.g. the lux). The decibel takes account of the relationship between sound pressure and intensity as perceived through the ear (auditory system) and so on. Measuring instruments often include those adjustments.

DOI: 10.1201/9781003401964-3

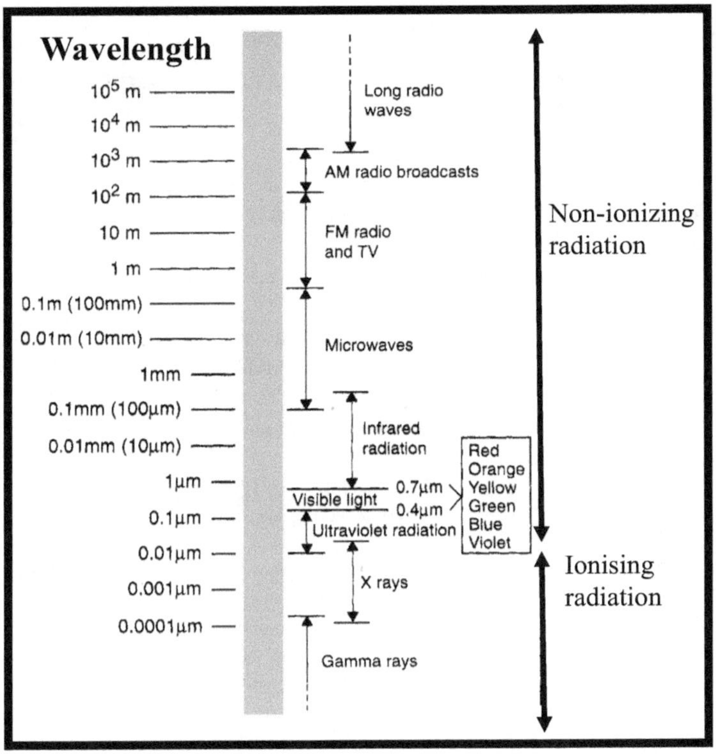

FIGURE 2.1 The electro-magnetic radiation spectrum. 0 in static fields (e.g. the earth) to <0.0001 µm, gamma rays. Ionizing radiation (<0.124 µm wavelength (mid-UV region)) takes or adds electrons to materials through which it passes. Visible light at 0.4 µm is seen as blue and at 0.7 µm it is seen as red if light levels are sufficient.

CHARACTERISTICS OF MEASURING INSTRUMENTS

The specification of a measuring instrument will define its performance at measuring the environment and must be sufficient to quantify the environment such that it can predict human response. Criteria include validity; accuracy; range; sensitivity; reliability; signal-to-noise ratio, specificity, response time, method of calibration and more. A specification for measuring instruments for environmental ergonomics is often provided in terms of regulations and standards to ensure that the environment is correctly represented (e.g. ISO 7726, 1998; IEC 61672-1, 2013; ISO/CIE 19476, 2014). See Tables 2.1 and 2.2.

A related point is that the sensory system of the body can also be considered in terms of similar criteria albeit from an organic perspective. It should also be noted that the person making the measurement will influence the outcome and should use an instrument correctly with an understanding of what is being measured and possible errors. All instruments will respond to the totality of the physical environment and an understanding of the variable of interest while minimising the effects of other

TABLE 2.1
Summary specification of instruments for measuring environmental components

Air temperature (ta)	
Measuring range	10–30 °C for comfort; –40 to 120 °C for stress
Accuracy	±0.5 °C required; ±0.2 °C desired for comfort
Response time	Shortest possible. To be specified
Comment	Protect sensor from radiation, ≥1 minute measurement desirable
Noise sound level meter dB(A)	
Measuring range	30–120 dB(A), up to 140 dB(lin)
Accuracy	Required, ±2 dB, desired, ±1 dB, for frequency range 20–20 kHz
Response time	125 ms as response time of human ear
Comment	Shield from low-frequency outdoor wind
Vibration accelerometers (awrms)	
Measuring range	Vibration frequency: 0.5 –80 Hz whole body; 0.1–0.5 Hz motion sickness; 0.063–1.0 Hz, high structures; 8–1000 Hz hand transmitted
	Vibration level: Depends upon frequency range and application. <0.01 ms^{-2} for building vibration. >2 ms^{-2} weighted acceleration equivalent to extremely uncomfortable in public transport. Significantly higher levels of vibration if unweighted as, for vibration acceleration, sensitivity to frequency generally reduces with increasing frequency
Accuracy	Within 10% of actual value depending upon application
Sensitivity	Depends upon accelerometer type (e.g. 1 mv/ms^{-2}) with acceptable signal-to-noise ratio to allow analysis
Response time	Well within measured frequencies and to allow valid sampling at no less than twice the highest frequency
Comment	See ISO 2631-1, 1997; and Griffin (1990) for a full discussion, including accelerometer type, natural frequency, damping, frequency range, phase errors, sensitivity including cross-axis sensitivity and environmental effects. Signal processing including sampling rate and range to match requirements for application
Light meter (horizontal Illuminance (lux)	
Measuring range	0.1–20000 lux depending upon application. Distinguish between artificial and natural light. Colour corrected for photopic (sometimes scotopic light). Usually a method for selecting range to match level
Comment	ISO/CIE 19476: 2014, Characterisation of the performance of illuminance metres and luminance meters provides definitions and quality indicators by which light metres are calibrated
Air quality (CO_2)	
Measuring	Determine the carbon dioxide and any change in carbon dioxide in the air (% or ppm)
Comment	Specification provided by the manufacturer of the instrument. Carbon dioxide level provides an indication of air quality both as a direct measure and as an indication of ventilation and air quality for other pollutants

TABLE 2.2
Example of commercial product specifications

Lawnmower

Noise levels measured according to 2000/14/EC at 1.6 m height, 1 m distance away are 83 dB(A) sound pressure level and 93 dB(A) sound power level.

Vibration levels (triaxial sum) determined according to EN 30335 are a_h *<2.5* ms^{-2}.

Luminaire – fluorescent tube

Rated average life is the lamp-burning hours to median life expectancy: 13000 hours.

Lamp power is the wattage of the lamp: 36.

Mean lumens is the average light output over a designated amount of time: 2500 lm.

Colour rendering index is an indication of a lamp's ability to render object colours in a normal natural way. It ranges from 0 to 100, a higher number indicating better colour appearance:

CRI = 72; colour temperature = 6200 K; light colour category = cool daylight.

variables is essential to obtain a valid measurement. Practice and some training may be required.

In practical application, simple errors such as talking when measuring sound, walking near an accelerometer when measuring vibration or shielding a thermometer, or light meter (with the body or the lid of the instrument case) from radiation that would normally be present at the position of interest, as well as general experimenter interference, are all common examples.

Validity is the degree to which what we measure reflects what we intend to measure. If we wish to measure light but the instrument only detects infrared radiation, then the measure will not be valid.

Accuracy assumes that there is an actual or correct value for the environmental variable the instrument is attempting to measure and gives an indication of how closely the instrument measure will be to the actual value. If people respond to 0.1 ms^{-1} of air velocity and the instrument only measures to 0.5 ms^{-1}, then it is not sufficiently accurate, but if it measures to 0.01 ms^{-1} then it is.

Range provides information about the lowest and highest value between which an acceptable measure of the environment can be made. It should span the range over which people are affected by the environment.

Sensitivity is related to the lowest level of energy an instrument can detect and how much of its measuring output it produces per unit of energy (in units appropriate to the environmental component). To obtain a useful measure of the physical environment for environmental assessment, a measuring instrument must have a sensitivity at least as good as that of a person, in the range of interest.

Reliability is the degree to which an instrument will indicate the same value when a measurement is repeated under identical conditions.

Signal-to-noise ratio is an index that demonstrates how detectable the measure of the desired variable is when compared with all other signals that make up background noise.

Specificity provides an indication of how the value measured by the instrument specifically relates to the measure of interest. An unshielded stationary thermometer, for example, used to measure air temperature will also be influenced by radiation.

Response time relates to how quickly an instrument responds when exposed to an environment. It should be sufficiently short to allow measurement when people occupy a changing environment.

Calibration is used to ensure that the instrument meets its specification. It usually involves comparison and adjustment of the instrument to ensure that it provides the same value as a high-quality standard instrument in a range of standard conditions. Simple calibration checks are often useful such as turning an accelerometer upside down to obtain a signal related to the acceleration due to gravity ($1\ g = 9.81\ ms^{-2}$) or placing identical temperature sensors in a stirred water bath with a standard calibrated thermometer to ensure that they give the same (correct) value. A pistonphone placed over a microphone will give a calibrated single-tone sound at a standard level (e.g. 1000 Hz at 94 dB).

Simple calibration checks before and after an environmental survey will be useful to test for changes over the period of measurement. An additional consideration is that the instrument may be influenced by the environmental conditions. Displays may not work in extreme cold or electronics may be influenced by heat for example. Calibration should then be conducted under conditions representative of the environments in which the instruments are used.

MEASURING THE SOCIAL ENVIRONMENT

Many environments are occupied by more than one person (not to mention social media) and that creates a social environment that can be influential in determining human response to the environment. Adaptive responses will be affected by the disposition of people to respond in a way influenced by cultural, social and peer pressure. Methods of measurement relate to the social sciences and involve quantitative, qualitative and discourse analysis (Wilson and Sharples, 2015a). It is important to characterise the social environment as part of an environmental ergonomics investigation.

A note on unethical practice is that the environmental ergonomist must be wary of suggestions that some groups of people tolerate or prefer what would usually be regarded as inferior environmental conditions, yet still maintain satisfaction. This is usually to the (financial) benefit of those not exposed to the environment, often encouraged and agreed by those exposed who feel, or are made to feel, special. Without clear evidence, however, this may not be the case. The additional argument that 'they don't expect anything better', must be viewed with similar caution. Even when satisfaction is expressed by those exposed, they may still prefer a 'better' environment.

An overlap between the measurement of the physical environment (this chapter) and the human response to the physical environment (Chapter 3), is the construction of equivalent contours. For a given level of sound, for example, the loudness at one frequency of sound may be perceived as different in loudness from a sound at a different frequency even though they are at the same physical level. There will, however, be a level of sound that will give an equivalent level of loudness. Equal loudness curves will then show levels across a frequency range where loudness will be equivalent. The Phon level is the equivalent loudness to that of a sound at a frequency of 1000Hz. An equivalent comfort contour for whole body vibration will be a curve or contour of vibration level over frequency where each point on the curve elicits equivalent comfort or discomfort.

If the relative level of the equivalent contour indicates 'sensitivity', then for complex environments with multiple frequency stimuli, the inverse of the sensitivity curve, normalised to 1.0 at the frequency of maximum sensitivity, provides a weighting function. These weighting functions can be represented by analogue or digital filters that can be used in transducers and signal processing so that the physical measure of the environment, when conditioned with a weighting function, provides a unit that is related to human perception and hence response to the physical environment.

The shape of the equivalent curve, and hence the weighting function, will depend upon the level of the component of the physical environment as well as the context and application. For light, where the radiation range is represented by wavelength, sensitivity to wavelength varies with light level and especially if we are concerned with colour vision (relatively high levels) or grey vision (low levels of light).

Equivalence between stimuli is determined through subjective assessment methods by comparison of sensitivity, comfort, annoyance, loudness, brightness and so on. As individuals differ in their sensitivity and contours change with level, idealised weighting functions are used in the measurement of the physical environment. Examples are provided in Figures 2.2–2.11.

The weighted values of measurements of the physical environment provide indices that relate to how people perceive and respond to the environment that is described in Chapter 3.

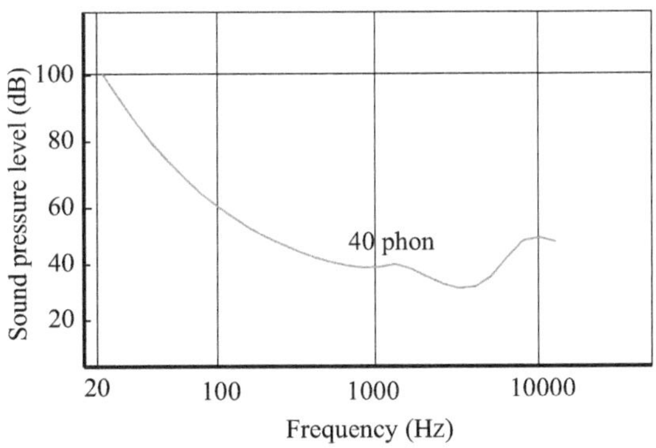

FIGURE 2.2 The 40-phon equal loudness curve for pure tones. (Compare inverse with Figure 2.3.)

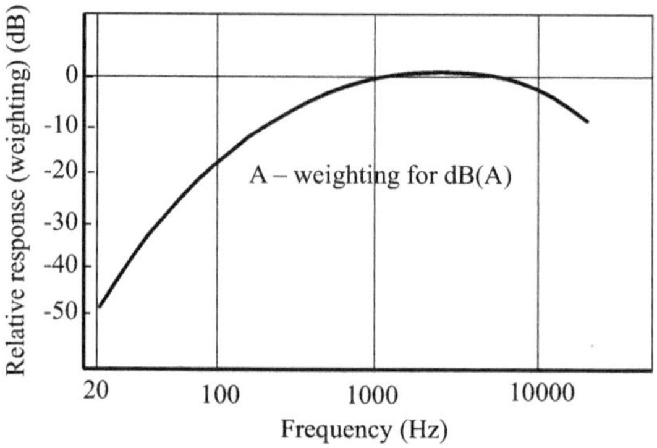

FIGURE 2.3 The A-weighting used to determine dB(A) and take account of the sensitivity of people to sound frequency. (Compare inverse with Figure 2.2.)

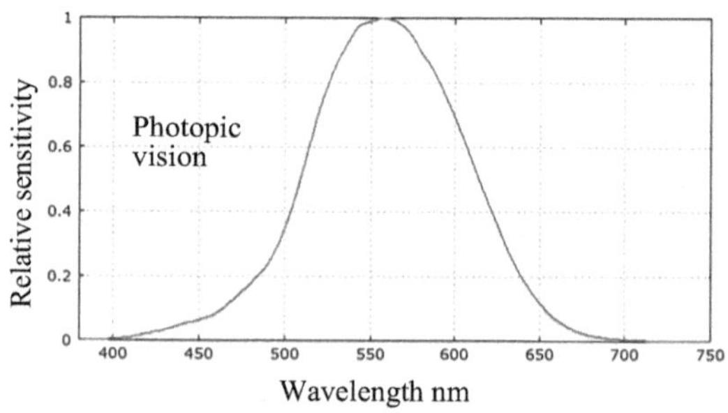

FIGURE 2.4 The 'weighting curve' for photopic vision (>3 cd m^{-2}).

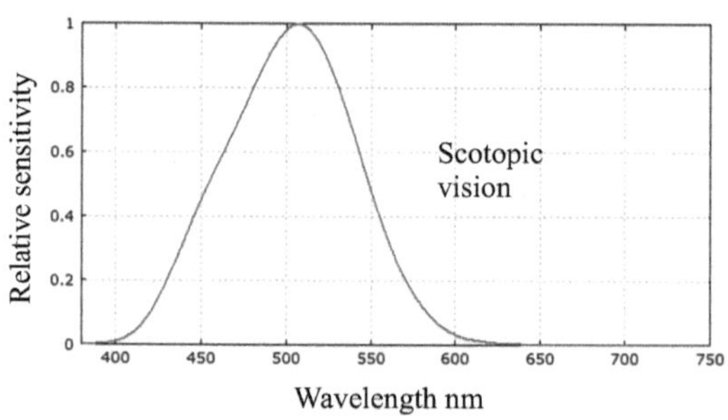

FIGURE 2.5 The 'weighting curve' for scotopic vision (<10^{-3} cd m^{-2}).

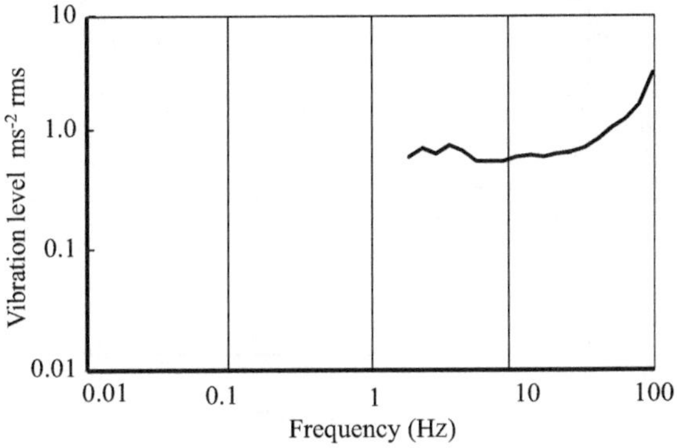

FIGURE 2.6 Equivalent comfort curve for z-axis whole-body vibration of a seated person. From Griffin et al. (1982). To be compared with the inverse of Figure 2.7.

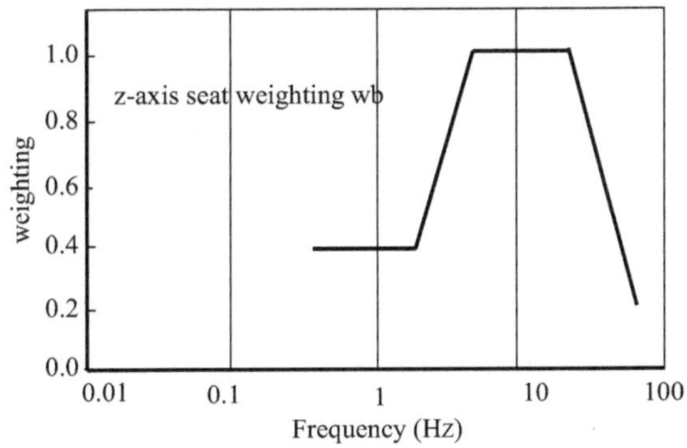

FIGURE 2.7 Idealised weighting curve for z-axis whole-body vibration for a seated person. The W_b curve has a value of 1.0 from 5 to 16 Hz; 0.4, from 0.5 to 2 Hz; *f*/5, from 2 to 5 Hz; and 16/*f* from 16 to 80 Hz. It can be compared with the inverse of the equivalent comfort curve provided in Figure 2.6.

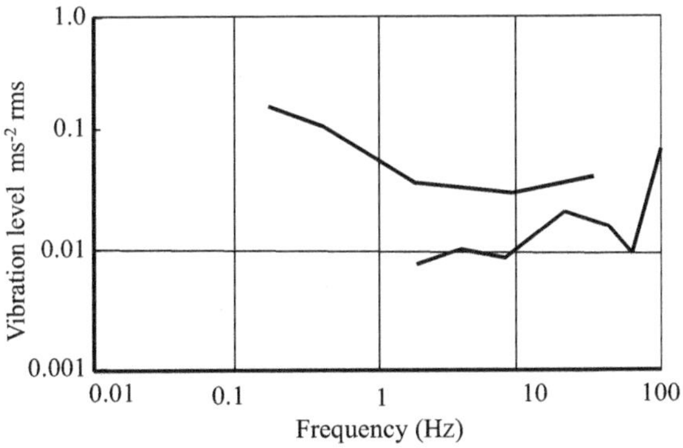

FIGURE 2.8 Absolute perception thresholds for z-axis whole-body vibration of seated persons. Data from Parsons and Griffin (1988) and Benson and Dilnot (1981). The inverse of the curves can be compared with the weighting function provided in Figure 2.9.

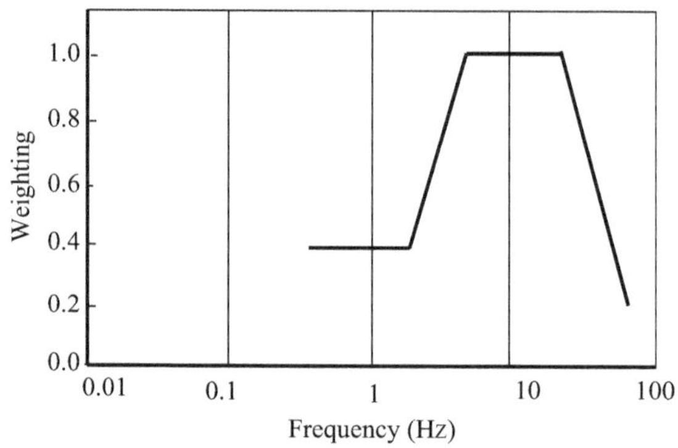

FIGURE 2.9 Idealised base curve showing relative sensitivity to z-axis whole body vibration of seated persons for levels close to absolute perception. Used with multiplying factors to provide limits for vibration in buildings (ISO 2631-1, 1997). Compare with inverse of Figure 2.8.

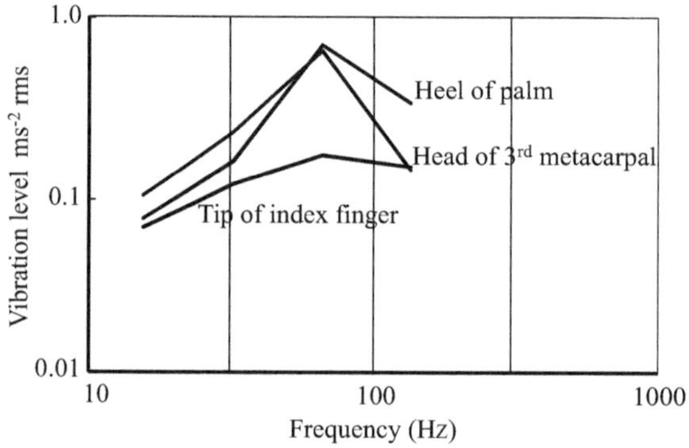

FIGURE 2.10 Absolute thresholds to vibration stimuli, for three areas of the hand (Data from Morioka, 1999). The inverse of the curves can be compared with Figure 2.11.

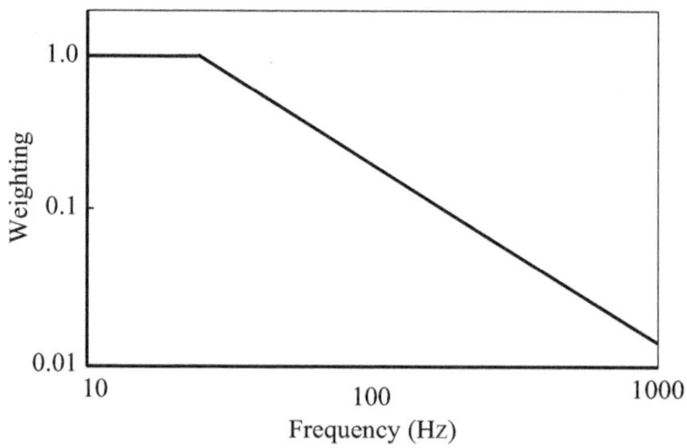

FIGURE 2.11 Weighting curve W_h for hand-transmitted vibration (ISO 8041, 1990). Can be compared with the inverse of the curves shown in Figure 2.10.

FURTHER READING

Wilson, J. R. and Sharples, S., 2015, Methods in the understanding of human factors. In Wilson, J. R. and Sharples, S., *Evaluation of human work* (4th Ed), CRC Press, Taylor & Francis Group, Boca Ratan, New York, Oxford, ISBN 978-1-4665-5961-5.

STANDARDS

IEC 61672-1, 2013, *Electroacoustics – Sound level meters – Part 1: Specifications*, International Electrotechnical Commission.

ISO 7726, 1998, *Thermal environments – Instruments and methods for measuring physical quantities*, ISO, Geneva.

ISO/CIE 19476, 2014, *Characterization of the performance of illuminance meters and luminance meters*, ISO, Geneva.

3 Measuring and Predicting Human Response to the Environment

MEASURING HUMAN RESPONSE

On detecting the environment and interpreting the signal or sensation, a person responds with an action, if disposed to do so, especially if the environment is causing an undesirable state. The perceived state of a person and actions taken can be measured with subjective, objective, behavioural and discourse methods (Wilson and Sharples, 2015a; Keats, 1971, Edwards, 1957) that include both quantitative and qualitative approaches (Hignett and Mc Dermott, 2015). They can also be predicted using computer models of human systems (Parsons, 2000).

SUBJECTIVE METHODS

People can report how they feel and how an environment is affecting them with some consistency. A powerful method for assessing environments is therefore to ask people about the effects of their environment and therefore use subjective methods.

Subjective assessment is a wide subject used extensively in applications to elicit information from medical assessment to police enquiries, market research, political trends and much more. It is considered here in the context of environmental ergonomics. It is implicitly assumed that a person receives information from a complex, dynamic and continually developing representation of the external world personal to them but with commonality as people have common characteristics and exist in similar environments.

One model (Kelly, 1955) is that a person develops personal constructs that structure their representation and that those constructs can be expressed semantically (hot, loud, bright, smelly, comfortable, painful, etc.). Subjective assessment in environmental ergonomics uses those semantic expressions to develop subjective scales that can be used for people to select a position on the scale that describes how they feel and that can be interpreted numerically.

It is possible to elicit semantic terms from a person by subjecting them to stimuli of interest and asking them to describe their feelings. The method of triads, for example, exposes people to three stimuli and divides them into one and a pair. The person then describes how the single stimulus differs or is similar to the pair. A new pair is chosen from the three and so on until the semantic terms have been elicited. When

DOI: 10.1201/9781003401964-4

semantic terms are derived and ratings obtained, correlation, factor analysis and multi-dimensional scaling techniques can be used to determine independent dimensions and scales (Shadbolt and Smart, 2015).

In practice, the semantic terms used in environmental ergonomics have been established through experience mostly independently by experts in that particular environmental component. Contextual factors for a particular environment may also influence the subjective scales used.

ISO 10551 (2019) provides suggestions for subjective scales to measure human responses to thermal, acoustic, light, vibration and air quality. ISO 28802 (2012) provides suggestions for how to measure the physical environment as well as subjective scales for use in the environmental ergonomics survey (Table 3.1). For all environmental components, five types of scale are proposed. These are sensation, comfort, preference, satisfaction and acceptability.

There are many others that could be included to cover aesthetics, pleasure, delight and so on. If an understanding of the subjective response is required, then asking how strongly a person agrees with their response and why they gave that response provide important information for the environmental ergonomist. This can lead to discourse analysis and a rich vein of information that would otherwise be missed.

Scales usually begin with qualifiers and ISO 28802 (2012) uses: not; slightly; '*the term' (e.g. uncomfortable)*; very and extremely to indicate intensity. They may also employ traditionally used terms such as the sensation scale for assessing thermal sensation (hot; warm; slightly warm; neutral; slightly cool; cool; cold) on a 7-point category scale or on a continuum.

Forced bipolar scales involve a choice (e.g. acceptable or not acceptable). Construction and use of subjective scales depend upon the aims and context of the environmental ergonomics survey and there are many options. International standards (e.g. ISO 28802, 2012) are translated into many languages by native speakers and should be considered in terms of the appropriate terms and culture for that language.

One advantage of standardisation is that if different studies use identical scales across the world, then results can be compared and knowledge is more easily enhanced. Parsons (2014) provides some translations for the thermal environment.

Suggestions for subjective scales used in an environmental ergonomics survey are provided in ISO 28802 (2012) (see Table 3.1) and fuller discussion is provided in Wilson and Sharples (2015a); Sinclair (2005); Edwards (1957); and Keats (1971) (see also Appendices 2–12).

Subjective assessment usually involves a person's selection of a position on a scale that can be represented by a number or a symbol. Thurstone (1927 – see Edwards, 1957) suggests that the intensity of the perception is not unique but that, for an identical stimulus, the conscious perception on an internal psychological scale will be distributed as a (normal) distribution of intensities. This allows consistency and possibly a more powerful interpretation of data.

The selected position can be regarded in a hierarchy of properties depending upon the assumptions made about the strength of the subjective judgement. A rating of 4 – 'very uncomfortable' is not twice the intensity of a rating of 2 – 'slightly uncomfortable' on the same scale but very uncomfortable is more uncomfortable than

TABLE 3.1
Physical measures and subjective terms and scales for use in environmental ergonomics surveys

Environment	Physical measure	Subjective term	Subjective scale(sl=slightly)
Thermal	air temperature	Sensation	hot,warm,sl warm,neutral,
	radiant temperature		sl cool, cool, cold
	air velocity	uncomfortable	Not..; sl …; uncom..;very..
	humidity	Stickiness	Not..; sl …; sticky; very..
	clothing	Draughty	Not..; sl …; Draughty;very..
	activity	Dryness	Not..; sl …; Dry; very…
		Preference	much warmer; warmer; sl warmer; no change; sl cooler; cooler; much cooler
		Satisfaction	Satisfied/Not Satisfied
		Acceptability	Acceptable/Not Acceptable
Acoustic	dB(A)	Annoying	Not..; sl …; annoying;very..
	dB(A) Leq	Preference	much quieter; quieter; sl quieter; no change
		Satisfied	Satisfied/Not Satisfied
		Acceptability	Acceptable/Not Acceptable
Vision and lighting	Horizontal illuminance (lux)	Discomfort	No..; sl …; discom..;much..
		Preference	much darker; darker; sl
			darker; no change; sl lighter; lighter; much lighter
		Satisfied	Satisfied/Not Satisfied
		Acceptability	Acceptable/Not Acceptable
Vibration	Weighted rms	Uncomfortable	Not…;sl…;fairly…;
			Uncom…;very..;extremely..
	VDV	Annoying	Not..; sl …; annoying;very..
		Satisfied	Satisfied/Not Satisfied
		Acceptability	Acceptable/Not Acceptable
Air quality	CO_2 level	Smelly	Not…; sl…;smelly;very…
		Satisfied	Satisfied/Not Satisfied
		Acceptability	Acceptable/Not Acceptable

slightly uncomfortable. It is reasonable to assume unique and 'greater than' relationships between the numbers but not interval and ratio properties.

Guilford (1954) identifies identity, rank order and additivity (numbers can be added, subtracted, multiplied and divided with validity) as the most important properties of numbers for measurement.

Stevens (1951) describes four types of scale in terms of valid relationships. A nominal scale is a category scale where symbols or numbers simply provide a unique identity but no hierarchy. For example, labels for yachts in a race or names of patrols in a scout group (e.g. owls, seagulls, etc.). One is not better or greater than the other but they are uniquely different and have identity.

An ordinal scale has rank order so also includes the relationship 'greater than' between the numbers (a progression from the ordinal scale would include an ordered metric where the intervals between the numbers can also be ranked).

An interval scale adds validity of intervals to the relationships between numbers. For example, the differences in temperature between 5 °C; 10 °C; 15 °C and 20 °C are all the same 5 °C; however, there is no absolute zero or ratio property. For example, 10 °C divided by 5 °C is not 2 °C, and 0 °C is when water freezes but we can measure −5 °C, etc.

The final scale is a ratio scale where there is an absolute zero and additivity applies as well as all of the other properties. Length is often given as an example of a ratio scale. For example, if we consider measurements of 0, 1, 2, 4, 7 and 12 cm. All measures are unique. 0 is absolute zero. In order of size, 12 has rank 1 and 0 has rank 6. If we rank the intervals in an ordered metric, 12 to 7 = 5, has rank 1; 7 to 4, rank 2; 4 to 2 rank 3; and 2 to 1 and 1 to 0 are a tie so average rank (4 and 5) is 4.5 for each interval. As a ratio scale, 12 − 1 = 11 cm; 12/4 = 3 cm; 7/4 = 1.75 cm; and so on. So for length nominal, ordinal (plus ordered metric), interval and ratio properties apply.

In environmental ergonomics, the practical point is how to come to valid decisions about how the environment is affecting people who occupy an environment from their subjective judgements. If we use continuous lines for the scales, then rational numbers can be achieved and ratio scales implied. However, which properties are then considered valid? Assumptions made should be justified (often by analysis) in terms of knowledge of the subjective phenomenon being considered.

In the presentation and analysis of data, if only categories can be assumed, then we are restricted to using frequency and mode values. If ordinal scales are assumed, then, in addition, median values and ranges will be possible. If interval values can be assumed, then parameters such as the mean and variance of numbers can be used. Ratio scales use the numbers with full mathematical validity but subjective judgements cannot usually be regarded as meeting the requirements of a ratio scale.

Assumptions are the prerogative of the person interpreting the data and may be correct or incorrect; however, they must be stated so that they can be challenged and alternative interpretations provided.

Presentation of subjective judgements over groups of people can be orientated in graphs, tables and more to provide an effective indication of the results of that part of an environmental ergonomics assessment. Decision-making using probability and statistical models can be divided into non-parametric statistics, where only nominal and ordinal properties of scales are assumed, and parametric statistics where at least interval properties are assumed. Siegel (1956) provides a full discussion and Field (2013) provides a comprehensive discussion with practical applications based on the universally accepted Statistical Package for the Social Sciences (SPSS).

Scales used for subjective assessment can be integrated into questionnaires. It is convenient to use electronic versions so that people who respond to questions can do so on mobile telephones, electronic tablets, i-pads and computers, including those used at the workplace in normal work. Paper-based questionnaires can also be used and examples of subjective scales and questionnaires are provided in Appendices 2–12.

Before conducting an environmental ergonomics survey using subjective assessment, it is useful (essential) to trial the questionnaire first with a similar cohort of respondents. This will not only ensure that the subjects will give meaningful responses but an analysis of the trial responses will clarify analysis and interpretation of the data to ensure that the actual study will meet its objectives.

OBJECTIVE METHODS

Objective methods in environmental ergonomics measure responses to the environment that are not based upon a person's perception, conscious action or subjective opinion. They require interpretation in terms of an underlying model (e.g. homeostasis) and require a consideration of specificity. Sweat rate, for example, may be an indication of a response to a thermal environment but it may also be a measure of anxiety and stress or both.

Instruments for taking objective measures require similar specificity as for measurements of the physical environment. They must not interfere significantly with what they are 'attempting' to measure (true for all instruments including subjective and behavioural measures)

Common objective methods used to determine human response to the environment are physiological and kinematic (related to movement). For thermal environments, sweat loss and body temperatures are relevant. Vibration uses accelerometers attached to the body to measure transmissibility and pupil diameter as well as eye movements can be used for assessing responses to visual environments. Chemical analysis of sweat may be useful for assessing the effects of air quality. Biological exposure indices (ACGIH, 2024) provide levels of chemicals in blood or urine, for example, above which would be considered harmful and unacceptable.

Heart rate; electrocardiography; electromyography; metabolic rate and breathing rate; and many more are all examples of objective measures. Which objective measures are used requires an a priori understanding of how the measures will be interpreted.

Because of the inconvenience, difficulty in interpretation and possible interference in taking objective measures; they are used sparingly in environmental ergonomics assessment. They can be useful in a follow-up assessment and, in particular, where effects on health are under investigation and subjective methods may become unreliable.

OBSERVATION: BEHAVIOURAL AND ADAPTIVE METHODS

An important and often determining response to an environment is for people to behave in a way that reduces undesirable conditions such as discomfort. The obvious response is to move away from the undesirable environment or to avoid what may become undesirable. Adjustments to the environment (opening or closing windows, adjusting thermostats, etc.) are also options. These are called adaptive responses.

As a trainee teacher, 'a tip' was to always 'publicly' check and make adjustments to the environment when starting a class, not only to provide optimum conditions, important for learning but also to demonstrate who was in charge!

These responses were always well known but often only mentioned in passing. They are now an integral part of environmental design and assessment and cover all environmental components and their integration into environmental ergonomics.

Adaptive opportunity (Baker and Standeven, 1997) is an important part of environmental design and allows an assessment of the environment in terms of what opportunities there are for people to behave to optimise their environment. It involves the opportunities provided and the ability, understanding and disposition of people to take them (Parsons, 2020).

In assessment, a checklist of observed behaviours or quantitative techniques such as observing the frequency of occupancy of different parts of a room (railway carriage, etc.) will provide information. Focus groups and interviews will provide fuller information on how and why people may provide behavioural and adaptive responses. Individual differences will be important. For example, a person with a physical disability may have a different range of adaptive opportunities from a person without a physical disability. Adaptive opportunities depend upon context and can be created as part of environmental design. A selection of adaptive opportunities for a range of environmental components is presented in Table 3.2.

TABLE 3.2
Examples of adaptive opportunities for a range of environmental components

<u>Too hot</u>

Principles: *Reduce air temperature; reduce radiant temperature; increase air velocity; reduce humidity; reduce clothing insulation; reduce activity; and minimise exposure time.*

Personal adaptive behaviour: *Move (go to the supermarket!); open or take off clothing; use cold water; use shades; avoid exposure (don't go on a journey; use cool part of the day); drink; avoid alcohol and drugs; change posture; select cool part of a room; avoid overconfidence; and so on.*

Technical adaptive behaviour: *Adjust thermostat; close blinds; use fans; open windows; use air conditioning; sit on and wear vapour-permeable materials; and so on.*

<u>Too cold</u>

Principles: *Increase air temperature; increase radiant temperature; reduce air velocity; reduce condensation; increase clothing insulation; increase activity; minimise exposure time; and so on.*

Personal adaptive behaviour: *Move (avoid draughts and cold windows, go to warmth and sun!); close or put on clothing; avoid exposure (don't go on a journey; use warm part of the day); avoid alcohol and drugs; change posture; select warm part of a room; be optimistic; avoid misery; be determined; and so on.*

Technical adaptive behaviour: *Adjust thermostat; open blinds; close windows; use heaters; and so on.*

<u>Acoustic</u>

Principles: *Reduce noise level at source; avoid sound frequencies and levels that cause annoyance and damage to health; and minimise exposure time.*

Personal adaptive behaviour: *Move to a quieter area; work at quiet part of the day; discuss noise reduction with others; complain; report hearing problems; and so on.*

Technical adaptive behaviour: *Wear ear protection; use noise cancelation device; construct barriers; close windows; close doors; and so on.*

<u>Visual field</u>

Principles: *Appropriate and interesting visual field, with horizontal illuminance at levels for task and avoidance of glare, flicker and reflections.*

Personal adaptive behaviour: *Move away from poor lighting environment; use daylight; orient position and posture to avoid glare and other sources of discomfort; and so on.*

Technical adaptive behaviour: *Adjust lighting; open or close blinds; use screens; make visual field interesting and inspiring (pictures and objects!); and so on.*

(*Continued*)

TABLE 3.2
(Continued)

<u>Vibration</u>

Principles: *Reduce vibration at source; reduce vibration level; avoid frequencies to which people are sensitive; and minimise exposure time.*

Personal adaptive behaviour: *Move away from vibration; reduce contact with vibration; reduce exposure time if possible (e.g. for vibrating tools); wear gloves; keep warm for hand vibration work; report health effects; and so on.*

Technical adaptive behaviour: *Use vibration isolation gloves and seats; use low vibration equipment; heated gloves and handles; and so on.*

<u>Air quality</u>

Principles: *Reduce odours at source; appropriate ventilation, air velocity, input and exhaust systems; and ensure fresh air to people.*

Personal adaptive behaviour: *Refuse to enter space if unacceptable; move away; complain of headaches and smells; and so on.*

Technical adaptive behaviour: *Open windows; introduce fresh air; mask smells with more pleasant odour (fresh air devices; perfumes); monitor* CO_2 *levels and respond if too high; and so on.*

DISCOURSE ANALYSIS

Discourse analysis provides a profound and rich source of information about how people view their environment. It involves people talking about their environment (in normal conversation, focus groups, personal interviews, group discussion, 'away days', etc.) and recording it for later analysis. It can be used in design and assessment.

Care must be taken in interpretation. Semantics used to describe responses to an environment can be elicited and, with simple techniques such as word count (wordl diagrams), priorities can be determined (e.g. if noise is mentioned many times, then the acoustic environment may require special consideration).

Discourse analysis is developing into a powerful tool for design. It should be recognised, however, that this is a holistic approach where the investigator and person or people engage in a social interaction.

This leads to the broader point relevant to all of the measures and methods discussed above. That is any environmental ergonomics assessment will change the nature of the social and personal disposition of those who occupy the environment as well as those involved in the assessment, and, although valid results can be obtained using the methods described, it would be naïve to ignore that influence.

My experience is that, 'all things being equal', in environmental ergonomics people are supportive and helpful providing altruistic information. It is a human disposition to take a position in any social interaction, however, and If, for example, it is to a person's advantage to demonstrate that an environment is satisfactory or unsatisfactory, then responses may be influenced.

People may 'wish to please' as well as give favourable results because someone is taking an interest in them or involvement in the survey becomes more interesting than their actual work. Distraction may be caused, which reduces time on work tasks and reduces performance. Because of sources of bias, measurement of objective responses as well as models of human response and regulations become of particular importance.

Attempts at purely behavioural responses may reduce 'investigator effects'. Determining cause and effect is then a matter of interpretation. The use of two-way mirrors, cameras or discreet observation can be useful and have been used for the assessment of classroom environments for school children (Humphreys, 1972; Wyon and Holmberg, 1972).

As an example of behavioural response, a school teacher once asked me to imagine the disruption and behaviour in her class when a leaking fume cupboard from the chemistry laboratory below her classroom leaked sulphur dioxide (smell of rotten eggs).

MODELLING HUMAN RESPONSE TO THE ENVIRONMENT

Modelling methods provide a simulation of human response to an environment. Parsons (2014) notes that a model will never be a perfect representation but can provide practical information, particularly where there may be a risk to health. Physical models can provide useful indices that relate to human response to the environment and computer simulations can contribute to computer-aided environmental design (Gallwey and O'Sullivan, 2005).

PHYSICAL MODELS

Physical models range from integrated instruments that respond to the variables to which people respond (microphone systems, wet and dry thermometer systems and more) to human-shaped manikins (whole-body or body part) adapted for simulating the effects of an environmental component on a person (male, female, adult, child, head, ear, nose, etc.).

COMPUTER MODELS

Computer models can represent a human body using rational methods involving passive (size, shape, properties of tissues, etc.) and dynamic (controlling systems and responses) systems and mathematical equations based upon the laws of physics.

For predicting responses to heat or cold, for example, dividing a body into idealised shapes such as cylinders and spheres and using the thermal properties of tissues (thermal conductivity, etc.) as well as thermoregulation to simulate homeostasis gives an example. Methods used include lumped parameter, finite element, finite difference and computational fluid dynamics (Parsons, 2014). Kinematic models for human response to vibration use mass, spring and damper systems and associated mathematics and computer simulation techniques to provide models of human response to vibration (Griffin, 1990; Mansfield 2004) and so on. Regulations and environmental guidelines and limits are often based upon models of human response to specific environmental components.

THE ENVIRONMENTAL INDEX

All environmental components (heat and cold, thermal comfort, noise, light, air quality, vibration) considered by the environmental ergonomist use indices to represent the integrated effects of relevant variables on human response. An environmental index is a single number that provides an indication of the human response to the environment. It is a representative value of the integration of the effects of all of the variables. It allows the specification and assessment of an environment in terms of a single number.

Although the individual values of variables in an environment may differ, if their combined effect produces the same index value, then we can assume that the effect on people will be equivalent.

For example, an index for thermal sensation could be a number related to the scale – 3 cold to +3 hot with an index value of +1 relating to slightly warm. Whatever the combination of values of relevant variables, if they integrate in the environmental index to provide a value of +1, then the predicted sensation is 'slightly warm'. An index can also be an equivalent physical value such as an acceleration value weighted with the sensitivity of the body to vibration frequency (see Chapter 2) or a value of the temperature of an instrument that responds as an integration of the effects of variables to which humans respond and so on. Much research has been conducted to provide useful indices for individual environmental components. These are provided in the following chapters and are often used to define regulations.

Little progress has been made in determining the effects of the interaction of environmental components (e.g. does the nature of lighting influence the effects of heat etc.) and there is currently no environmental index for total environments. Such an index could be a mark of environmental quality.

A simple approach would be to derive index values for individual components and integrate them in a logical way. For example, the most severe component could be regarded as the index for the total environment. A root mean square value (or other exponent) or a masking method, would allow a contribution from all environmental components.

Another complementary approach would be to convert indices for each environmental component to a common ubiquitous subjective scale (comfort, acceptability, etc.). The ratings for the scale values for each component on the common scale could then be integrated as described earlier to provide a final rating for the environment. It will be difficult to demonstrate cause and effect but such an approach will have value in terms of utility with face validity and reliability, if methods are defined.

This chapter has provided fundamental principles and a general introduction to methods used in environmental ergonomics. It is emphasised that the methods described can be used in combination to provide a full assessment. The following chapters provide details in terms of specific environmental components. Chapter 24 provides a case study that uses information from other chapters to provide an integrated environmental ergonomics survey.

FURTHER READING

Wilson, J. R. and Sharples, S., 2015, Evaluation of human work (4th Ed), CRC Press, Taylor & Francis Group, Boca Ratan, New York, Oxford, ISBN 978-1-4665-5961-5

Parsons, K. C., 2000, Environmental ergonomics: A review of principles, methods and computer models, Applied Ergonomics

STANDARDS

ISO 9886, 2004, Evaluation of thermal strain by physiological measurements, ISO, Geneva

ISO 10551, 2019, Ergonomics of the physical environment-Subjective judgement scales for assessing physical environments, ISO, Geneva

ISO 28802, 2012, Ergonomics of the physical environment – Assessment of environments by means of an environmental survey involving physical measurements of the environment and subjective responses of people, ISO, Geneva.

Part II

Thermal Environments

Part 2 includes four chapters. It provides principles behind the measurement and assessment of the thermal environment (Chapter 4); and demonstrates methods in case studies concerned with thermal comfort (Chapter 5); heat stress (Chapter 6); and cold stress (Chapter 7).

People burn food in oxygen in the cells of the body and produce heat measured in Joules (J). The average kinetic energy in a body is related to its temperature and Fahrenheit, Celsius, and Kelvin have produced scales. The body 'attempts' to maintain an internal body temperature of around 37 °C and has a system of human thermoregulation that defends that position involving vasodilation and sweating when too hot and vasoconstriction and shivering when too cold. Behavioural and adaptive methods contribute to psychological responses where thermal comfort is preserved in a holistic system of psychophysiological thermoregulation (Figure P2). This forms the fundamental model for people in thermal environments used in an environmental ergonomics survey of the thermal environment.

Chapter 4 considers the thermal environment in the context of environmental ergonomics. Despite much research and understanding, people still die of heat and cold and thermal conditions are a major source of discomfort and dissatisfaction. The interaction of air temperature, radiant temperature, air velocity, humidity, clothing and activity defines thermal stress. Definitions, units and methods of measurement are provided. The thermal audit demonstrates the body heat method of analysis. Current international standards are presented.

Chapter 5 presents a case study concerned with thermal comfort in an open-plan office. Fred is the manager of the office and contacted Sheila an environmental ergonomist who, with her team, conducted the survey. They measured the thermal environment and the human response using subjective and observational methods

DOI: 10.1201/9781003401964-5

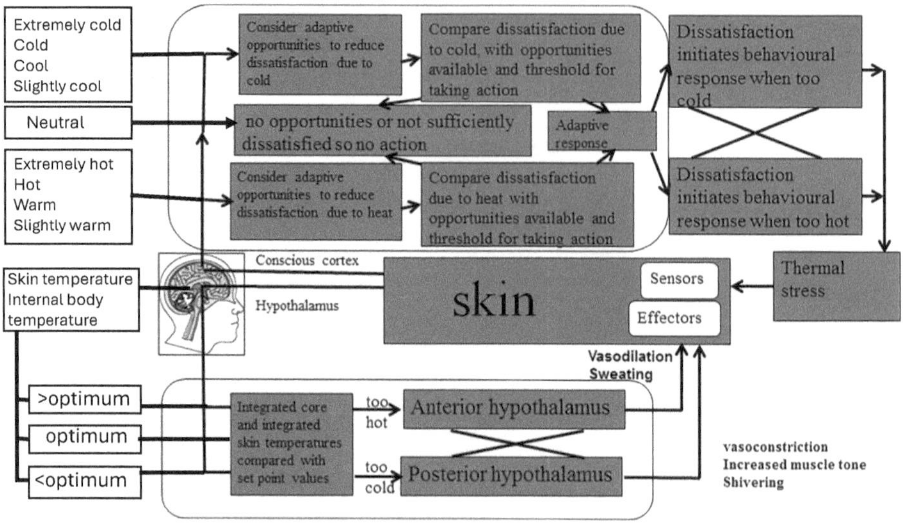

FIGURE P2 Psychophysiological system of human thermoregulation (thermal stress depends upon the interaction of air temperature, radiant temperature, air velocity, humidity, clothing and activity).

and predicted responses using the PMV/PPD indices and local thermal discomfort using ISO 7730 (2024). The predictions are compared with accepted values for offices. Recommendations include a redesign of the ventilation system, an increase in air temperature, a more flexible policy on clothing and a consideration of how adaptive opportunities can be introduced.

Chapter 6 considers the avoidance of heat casualties in special forces selection trials. A case study is presented to demonstrate methods and standards for the assessment of heat stress. Sebastian is head of selection and training for special forces personnel. There have been deaths in selection trials due to heat stress and procedures are required to prevent them. David has been contacted by Sebastian and has the remit to set up safe procedures. He contacted Thomas and his team of environmental ergonomists at the University of Old Hartley. Sebastian tested Thomas with an existing case of a highly motivated individual who set off on a mission in the heat, became confused and was captured. Thomas used the WBGT index, the PHS index and a model of human thermoregulation to predict an accurate outcome that gave confidence to Sebastian in the environmental ergonomics investigation and subsequent advice.

The mountainous course for the selection trails, its climate and other relevant details were supplied to Thomas. Advice was provided on instruments for measuring heat stress and methods for estimating metabolic heat production and clothing properties. The UTCI and WBGT indices were recommended as indicators of likely unacceptable heat strain. Internal body temperature and heart rate were recommended as physiological measures that provide limits for heat strain. Recommended methods for determining levels of heat strain included the UTCI climate alert index; ISO 7243 (2017) and the WBGT index; ISO 7933 (2023) and the PHS index; and the 2-node model of human thermoregulation. Further general advice was provided and it was recommended that David and his team be trained and implement a system of procedures and contact Thomas when this first stage had been completed.

Tom contacted Seo-ah to conduct an environmental ergonomics survey of a cold store he managed as a supply depot for a large supermarket chain (Chapter 7). Seo-ah measured the environment in the cold store and changing rooms and Lisa measured the subjective responses of workers. The clothing insulation required (IREQ) and the chilling temperature (tch) were calculated from the measurements according to ISO 11079 (2007). It was recommended that clothing insulation should be at around 2.3 clo with worker discretion; fans should be turned off when work was being conducted; waiting periods in the cold (−23 °C) should be avoided; and maintenance work involving fine dexterity will require further consideration.

4 Environmental Ergonomics and the Assessment of the Thermal Environment

HUMAN RESPONSE TO THERMAL ENVIRONMENTS

Despite many years of research into the human responses to the thermal environment, environmental ergonomics surveys continue to find thermal discomfort a major source of dissatisfaction and people still die of heat stress and cold stress as well as damaging their skin when in contact with very hot or cold surfaces. The thermal environment is therefore an essential part of any environmental ergonomics assessment.

Whether we are concerned with heat stress, cold stress or thermal comfort, a person's thermal environment is made up of six basic variables that are usually represented as parameters. These are air temperature, radiant temperature, air velocity, humidity and the personal factors of activity level and clothing. For environments not in air (water, space, etc.), modifications are required but the principles of assessment remain the same.

An important point is that it is the combined interaction of the six basic parameters that determines human response which includes physiological responses as well any behavioural responses that a person uses in an attempt to occupy a more desirable environment. A seventh factor that is important for environmental assessment is the opportunity to behave in a way that will reduce discomfort, a threat to survival or enhance pleasure. This is referred to as the adaptive opportunity (Baker and Standeven, 1997).

HUMAN RESPONSE TO THERMAL ENVIRONMENTS – A WORKING MODEL FOR ENVIRONMENTAL ERGONOMICS

People generate heat in their bodies by burning food, which they eat, in oxygen, which they breathe. They use this (metabolic) heat to maintain homeostasis at an internal body temperature of around 37 °C. The rest of the heat they remove from the body by heat exchange, through any clothing, with the environment. If the environment does not allow sufficient heat to be lost, body temperatures rise and the body suffers heat stress with possible damage to health and eventually

DOI: 10.1201/9781003401964-6

death. If the environment is such that too much heat is lost from the body and its temperature falls, then it suffers cold stress with possible damage to health and eventually death.

If the body responds to regulate heat exchange to maintain internal body temperature, such as by sweating or lowering skin temperature by removing blood from extremities, then this is termed thermal strain and the body may suffer discomfort and loss in performance. If there is no thermal strain, then it is likely that thermal comfort will be achieved. Paradoxically, if a stimulating and pleasant environment is to be achieved, that is beyond thermal comfort and satisfaction, then a dynamic environment involving some thermal strain, will be required.

In addition to whole-body responses, there are local responses of the body to thermal stimuli causing draughts in cool moving air, asymmetric radiation effects, thermal gradients and more. Contact with solid surfaces and liquids can lead to skin damage such as burns and frostbite and feelings of pain, hot, warm, cool, cold and thermal comfort and discomfort.

Comprehensive treatments of human response to the thermal environment are provided in McIntyre (1980), Parsons (2014) and Wilson and Sharples (2015a). For a focus on heat stress, see Parsons (2019), thermal comfort (Parsons, 2020) and cold stress (Parsons, 2021).

THE HUMAN THERMAL ENVIRONMENT

Energy, Heat and Temperature

All matter can be considered to consist of particles (atoms, molecules, etc.) that move due to their energy content, and, if there is no energy, they stop. Energy due to movement is termed kinetic energy, the amount of which is related to mass and velocity. The average kinetic energy of particles in a body is related to its heat energy and is its temperature. When there is no energy, the temperature is said to be absolute zero, that is zero degrees Kelvin (K).

In everyday use, temperature is often reported as a relative scale. The Celsius scale takes zero (0 °C) as the freezing point of water and 100 °C as the boiling point of water (under standard conditions) and hence is sometimes called centigrade. Absolute zero (0 K) is –273.15 °C and a 1 K difference in temperature is the same as a 1 °C difference in temperature. Therefore, for a room at 20 °C air temperature, it is 293.15 K.

The Fahrenheit scale takes its zero point at the freezing point of salted brine (some say it was originally the lowest temperature experienced in Fahrenheit's hometown in Poland and it is also the freezing point of a solution of alcohol) and the freezing point of water is 32 °F. There are 180° between the freezing point of water and its boiling point so the boiling point of water is 212 °F.

To convert degrees Celsius to degrees Fahrenheit, multiply by 180/100 (= 9/5 so divide by 5 and multiply by 9) and add 32. To convert degrees Fahrenheit to degrees Celsius subtract 32, divide by 9 and multiply by 5 (i.e. multiply by 100/180 = 5/9). Therefore, an internal body temperature of 37 °C corresponds to 98.6 °F, etc.

MEASURING THE THERMAL ENVIRONMENT

AIR TEMPERATURE

In environmental ergonomics, the air temperature (ta) is the temperature of the air that drives heat exchange by convection between the body and the environment.

It is measured with a thermometer with properties according to ISO 7726 (1998) – see Table 4.1. The thermometer is placed in a position to provide the value according to the definition above. It should not be too close to the body as the air will be influenced by the body and not too far away such that it is not representative of the air in the environment that drives heat exchange. It should be quantified in both space and time if there is significant variation. Precautions to avoid the influence of other variables, particularly radiation, include using a silver surface for the sensor, shielding the sensor (but still allowing air flow) and passing air across the static sensor or rapidly moving the sensor through the air (as in a whirling hygrometer).

Instruments for measuring air temperature include thermocouples, thermistors, platinum resistance thermometers, semiconductor junctions and mercury in glass thermometers (not now recommended in case of breakage). Parsons (2014) and Pandolf et al. (1988) provide details. Thermistors used with data loggers (sometimes small self-contained button-type systems) or telemetry systems with remote signal reception and recording provide practical solutions for measurement over time and throughout a space. The electrical resistance of the thermistor changes with temperature and the relationship between the two is used to provide temperature for a measured resistance. The use of mobile communication systems (e.g. mobile 'phones) and the internet provide convenient and effective recording and analysis systems for the whole range of instruments and variables, including air temperature, for measuring the environment.

TABLE 4.1
Specification of instruments for measuring air temperature (t_a) (c = comfort s = stress)

Measuring range (°C or stated)	Accuracy (°C or stated)	Response time (min)	Comments
Air Temperature (*ta*)			
10 to 40 (c)	Required: ± 0.5 Desirable: ± 0.2	shortest possible	Response time assumes measures taken in air.
–40 to 120 (s)	Required:	shortest possible	
–40 to 0:	$\pm(0.5+0.01\|t_a\|)$		
0 to 50	± 0.5		
50 to 120	$\pm(0.5+0.04(t_a-50))$		
	Desirable: required/2		

ISO 7726 (2001) suggests a range of 10 °C to 40 °C for comfort assessment extended to –40 °C for the assessment of cold stress and to 120 °C for the assessment of heat stress. In practical application, an accuracy of at least ±0.5 °C should be achieved and the response time of the sensor should be as short as possible. A mean value over at least 1 minute is desirable.

Radiant Temperature

All bodies above a temperature of absolute zero emit and absorb radiation and exchange radiation with each other, so in an environment with a number of objects, there is a radiant field. There is a net flow of radiation between two bodies from the body at the higher temperature to the one at the lower temperature and the rate of heat exchange is proportional to the difference in the fourth power of their absolute temperatures.

The constant of proportionality is the Stephan-Boltzman constant (5.67×10^{-8} W m^{-2} K^{-4}). Thermal radiation is part of the electromagnetic radiation spectrum (Figure 2.1). It can travel through a vacuum such as space and will have a peak wavelength that is relatively short at high temperatures such as from the sun (with a surface temperature of approximately 5700 K) and at longer wavelength at lower temperatures such as from walls in a room at 293.15 K or 20 °C (Wein's law).

The radiant heat exchange between a person and his or her surroundings is therefore related to the difference in the (fourth powers of) surface temperatures of the person and the integration of the temperatures in the surroundings. Moderating factors include projected surface areas in different directions in a three-dimensional space as well as reflectivity and absorptivity of materials to radiation which depend upon wavelength. People wearing white clothing outside in the sun reflect the short wavelength radiation but in longer wavelength indoor radiation white clothing has no 'advantage' in hot conditions. A theoretical black body absorbs and emits all wavelengths of radiation.

Mean radiant temperature (tr) at a point in a radiant field represents the mean radiation averaged over all directions at that point. It is the temperature of a uniform enclosure with which a small black sphere at the test point would have the same radiation exchange as it does with the 'real' environment; that is, the temperature of a uniform hypothetical environment in which the temperature of the walls equals the air temperature and gives the same heat exchange at that point by radiation as in the actual environment.

The mean radiant temperature of an object has a similar definition but for the object and not a point or sphere. It includes the characteristics of the object, for example, its projected surface areas in all directions. If the object is the human body, then body shape and posture will be important.

Plane radiant temperature (tpr) is the radiant temperature in a particular direction. It is the uniform temperature of an enclosure where the radiance on one side of a plane element is the same as in the (non-uniform) actual environment. It can be thought of as representing the radiation exchange for each of the six faces of a cube at a point; and for the human body, the values can be weighted depending upon posture and projected area to provide an estimate of mean radiant temperature. ISO 7726 (2001) provides weighted values for different body postures (sitting, standing, etc.).

Radiant temperature asymmetry is the difference between the plane radiant temperatures of the two opposite sides of a small plane element. It is useful in an environmental survey where radiation is directional and its effects cannot be represented by mean radiant temperature. Suggested values related to dissatisfaction are provided in ISO 7730 (2024).

Instruments for measuring radiant temperature include a black globe thermometer, a heated globe thermometer, a shielded globe thermometer, a net radiometer and a pyranometer.

For the environmental ergonomics survey, the radiant environment can be quantified using a black sphere with a thermometer at its centre (black globe thermometer) and giving globe temperature (t_g) at the position of the person (to give mean radiant temperature at a point when corrected for air velocity and air temperature; Bedford and Warner, 1934).

For natural convection $v < 0.15$ ms^{-1}, t_r (°C) is given by

$$t_r = \left[\left(t_g + 273\right)^4 + \frac{0.25 \times 10^8}{\varepsilon} \left(\frac{\left|t_g - t_a\right|}{d} \right)^{\frac{1}{4}} \times \left(t_g - t_a\right) \right]^{0.25} - 273 \tag{4.1}$$

For forced convection $v \geq 0.15$ ms^{-1}

$$t_r = \left(\left(t_g + 273\right)^4 + \frac{1.1 \times 10^8 v^{0.6}}{\varepsilon d^{0.4}} \times \left(t_g - t_a\right) \right)^{0.25} - 273 \tag{4.2}$$

Therefore, the larger the globe (diameter d), the lower the air velocity (v) and the closer t_a is to t_g, the closer t_r is to t_g. If $t_g = t_a$, then $t_r = t_g$ and there is an exchange but no net heat gain or loss by radiation. If $t_g > t_a$, then there is a net heat gain by radiation from the environment, and if $t_g < t_a$ there is a net heat loss. A standard black globe of diameter $d = 0.15$ m and emissivity $\varepsilon = 0.95$ (often (but not essentially) made of copper to give a relatively quick response time – still up to 20 minutes) is often used. Smaller globes are sometimes more convenient to use but care must be taken as errors in the measurement of variables used for correction (particularly air velocity) can become significant.

Plane radiant temperature can be considered if radiation appears dominant in one or more directions and a pyranometer can be used where solar radiation is present. More detail is provided in Pandolf et al. (1988), ISO 7726 (2001) and Parsons (2014). The specification of instruments for measuring radiant temperature is provided in Table 4.2.

TABLE 4.2
Specification of instruments for measuring radiant temperature (c = comfort; s = stress)

Measuring range (°C or stated)	Accuracy (°C or stated)	Response time (min)	Comments
Mean Radiant Temperature (*tr*)			
10 to 40 (c)	Required: ± 2 Desirable: ± 0.2	shortest possible	Accuracy less for globe method
–40 to 150 (s)	Required:	shortest possible	State accuracy if required is not achieved
–40 to 0:	$\pm(5+0.02\vert t_a\vert)$		
0 to 50	±5		
50 to 150	$\pm(5+0.08(t_r-50))$ Desirable: required/2		
Directional radiaton (rd)			
–35 to +35 Wm^{-2} (c)	Required: ±5 Wm^{-2} Desirable: ±5 Wm^{-2}	≤1 s	Accuracy depends
–300 to –100 Wm^{-2} (s)	Required: 10% Desirable 5%		
–100 to +1000 Wm^{-2}	Required: ±5 Wm^{-2} Desirable ±5 Wm^{-2}		
> +1000 Wm^{-2} (s)	Required: 10% Desirable 5%		
Plane radiant Temperature (tpr)			
0 to +50 (c)	Required: ±(0.6 + 0.05|tpr|) Desirable: ±(0.2 + 0.04|tpr|)		Guaranteed for at least
–60 to +200 (s)	Required: ±(1.2 + 0.02|tpr|) Desirable: ±(0.6 + 0.02|tpr|)		$\vert tpr - t_a\vert < 10$ °C (c) $\vert tpr - t_a\vert < 50$ °C (s)
Surface Temperature (ts)			
0 to +50 (c)	Required: ± (0.6 + 0.01|ts|) Desirable: ± (0.15 + 0.002|ts|)	≤1 min	Accuracy depends upon contact
–50 to +200 (s)	Required: ± (0.6 + 0.01|ts|) Desirable: ± (0.15 + 0.002|ts|)		Pressure
Globe Temperature (t_g)			
0 to +50 (c)	Required: $\pm(0.6 + 0.01\vert t_g\vert)$ Desirable: $\pm(0.15 + 0.002\vert t_g\vert)$	≤30 min	For a globe 150 mm in
–50 to +200 (s)	Required: $\pm(0.6 + 0.01\vert t_g\vert)$ Desirable: $\pm(0.15 + 0.002\vert t_g)$		diameter

A pyranometer measures radiation from the sun over wavelengths that are detected by people and provides a measure of the energy (usually in Wm^{-2}). The energy from the sun at the surface of the atmosphere varies but is around 1370 Wm^{-2} and the maximum energy in full sun on the earth's surface can be up to1000 Wm^{-2}. When out of the sun, this can be reduced to zero and can become negative at night, especially in a clear sky. Shade provides shielding from the direct sun and can reduce heat stress and be used as an adaptive opportunity to achieve thermal comfort. It should be remembered that, even in shade, diffuse radiation and reflections may still contribute to the radiation environment.

Air Velocity

In environmental ergonomics, air velocity is the speed and direction of air across the body. It can be measured through time and space and is often represented as an average or steady state value. Representative values of whole-body average air velocity are used to predict heat loss and human response and local air velocities are often used to predict local heating, cooling and discomfort caused by draught. Turbulence intensity is represented by dividing the standard deviation of the air speed by its mean and expressed as a percentage and is particularly relevant to predicting draughts. Relative air velocity is when the body moves and the vector sum of environmental and body movement provides the resultant air velocity across the body.

Instruments for measuring air velocity (anemometers) include a hot wire anemometer; heated sphere anemometer; Kata thermometer; cup anemometer; vein anemometer; smoke; and bubbles. Anemometers that involve mechanical movement (e.g. cup and vein) will be useful for measuring relatively high air movements (e.g. in ventilation shafts and outdoors) but will be limited in low air velocity environments such as in offices as people can detect air movements as low as 0.1 ms^{-1}. That is, below the working lower threshold of some instruments (e.g. due to inertia).

Hot wire anemometers use the electrical current required to maintain the temperature of a hot wire, in combination with air temperature to estimate air velocity. They are sensitive but have directional properties that must be taken into account. Hot sphere anemometers use a similar principle but provide an integrated, omnidirectional air velocity. Smoke and soap bubbles provide a visualisation of air velocity but may contaminate the environment. When used with photographic techniques (e.g. mobile telephones), they can provide information on speed and direction as well as distribution profiles. I have found that creating bubbles (using instruments but also a children's toy waved (not blown by mouth) through the air, or a variety of ingenious systems from bubble guns and bubble fans to bubble carts, provides an excellent starting point for assessing the air velocity at a workplace. A specification of requirements for anemometer use in an environmental ergonomics survey is provided in ISO 7726 (2001). The specification of instruments for measuring air velocity is provided in Table 4.3.

TABLE 4.3
Specification of instruments for measuring air velocity (va) (c = comfort; s = stress)

Measuring range (ms^{-1})	Accuracy (ms^{-1})	Response time (sec)[a]	Comments
0.05 to 1.0 (c)	Required: ±(0.1 + 0.05 |va|)	≤2	Ranges match
	Desirable: ±(0.05 + 0.05 |va|)	≤1	accuracy
0.1 to 20 (s)	Required: ±(0.1 + 0.05 |va|)	≤1	
	Desirable: ±(0.1 + 0.03 |va|)	≤2	

[a] Response time for turbulence ≤0.2 sec and of sufficient frequency.

Humidity

Humidity is the amount of water vapour held in the air (absolute humidity). It influences human response to the thermal environment particularly in hot environments where it is significant in determining heat loss by the evaporation of sweat.

The total (maximum) amount of water vapour that can be held in the air depends upon the air temperature (more water vapour can be held at higher temperatures) and at that level, the air is said to be saturated. Above saturation, water condenses out of the air and if air is cooled, by coming into contact with cold windows for example, the saturation temperature will fall, the amount of water vapour in the air may become above the saturation level and water will condense (e.g. clouding car windows).

The temperature at which water vapour condenses out of the air is termed the dew point. When using chilled beams and ceilings for cooling in offices, care must be taken not to reduce the temperature of the ceiling (using a system of cold water pipes, for example) below dew point as 'office rain' will fall.

The water vapour in the air exerts a pressure that is termed the partial vapour pressure (in Pascals where 1 Pa = 1 Nm^{-2}). The relative humidity (rh or Ø) is the ratio of the amount of water vapour in the air to the total amount of water vapour that can be held in the air at that temperature. It is often given as the partial vapour pressure in the air divided by the saturated vapour pressure at that temperature and represented as a percentage.

An instrument for measuring humidity is called a hygrometer (sometimes called a psychrometer). Water vapour can cause a change in electrical capacitance that on calibration, and with air temperature, can give a direct measure of humidity. The change in the length of a (horse) hair provides a simple measure (often linked to a pen chart to give a continuous indication of humidity in the workplace), and use of a cooling element to identify when water condenses out of the air (often using optical means) can provide the dew point.

A psychrometric wet bulb thermometer provides maximum cooling by evaporation (caused by air movement). The temperature of the wet sensor is reduced by evaporation to below air temperature up to saturated vapour pressure where there is no further net loss of heat by evaporation. A whirling hygrometer rotates a frame containing a wet and dry bulb thermometer to provide dry bulb temperature (with

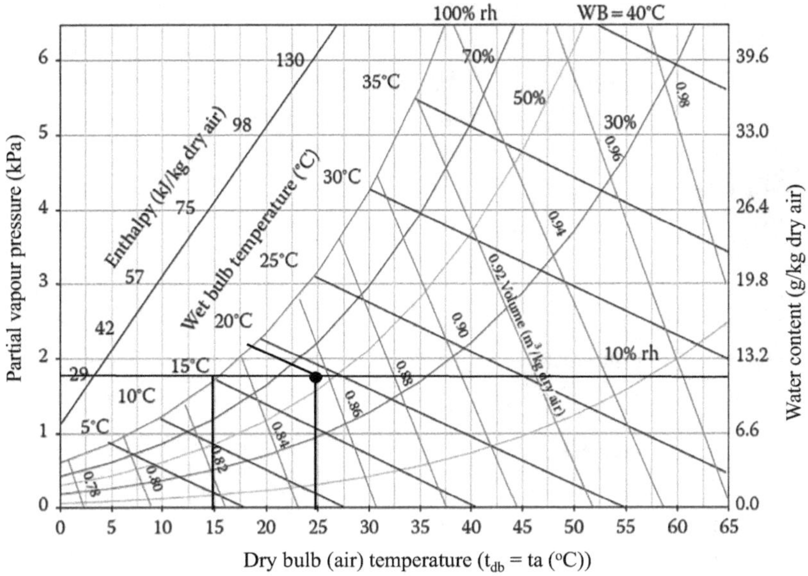

FIGURE 4.1 Psychrometric chart. The example plots a psychrometric wet bulb temperature of 18.5 °C and a dry bulb temperature (air temperature) of 25 °C. The intersection provides a point on the chart giving relative humidity as 55%; dew point, 15 °C; water content, 10 g/ kg of dry air; and partial vapour pressure of 1.7 kPa.

negligible effects of radiation) and psychrometric wet bulb temperature. The whirling must be fast to provide maximum cooling and stable lowest temperature (or a fan system used such as in the Assman psychrometer). A psychrometric chart (Figure 4.1) allows humidity to be determined in its various forms as well as the heat loss and gain by mass and heat transfer of air from one state of temperature and humidity to another (enthalpy). Equations that provide approximations for determining related representations of humidity are provided in Parsons (2014). A specification of instruments for measuring humidity is provided in Table 4.4.

INTEGRATING INSTRUMENTS

Bedford 1940), Chrenko (1974), see also Ellis et al. (1972), and Parsons (2014) describe a practical system for measuring thermal conditions on Royal Navy ships in the United Kingdom. This was called the Admiralty Box. Although the instruments were not powered and well before the computer and internet age, they are still of great practical use, especially in training where the instruments can be related to fundamental principles (Figure 4.2).

Although more sophisticated integrating systems are available involving telemetry, the internet, local storage and analysis and more, for example, the B&K indoor climate analyser is well documented (Olesen, 1985). For the thermal environment, as for all environmental measurements, the environmental ergonomist must be aware of the principles of measurement and of what is being measured.

TABLE 4.4
Specification of instruments for measuring humidity (c = comfort; s = stress)

Measuring range (°C or stated)	Accuracy (°C or stated)	Response time (min)	Comments
Partial vapour Pressure (Pa)			
0.5 to 3kPa (c)	± 0.15 kPa	shortest possible	guaranteed for
0.5 to 6 kPa (s)	± 0.15 kPa		$\|t_r - t_a\|$ at least 10 °C(c); 20 °C(s)
Dew point Temperature (t_{Dew})			
–5 to +28 (c)	Required: 0.2		
	Desirable: 0.1		
–5 to +50 (s)	Required: 0.5		
	Desirable: 0.2		
Relative Humidity (RH)			
20% to 80% (c)	Required: 3%	≤3	Range for class (s)
(10 to 35 (°C))	Desirable: 2%		is best available
5% to 95% (s)	Required: 3%		
	Desirable: 2%		
Psychrometric Wet bulb Temperature (t_w)			
5 °C to 40 °C (c)	Required: ± (0.6 + 0.1 $\|t_w\|$)	≤1	Response time in air
	Desirable: ± (0.05 + 0.002$\|t_w\|$)		
Natural Wet bulb Temperature (t_{wb})			
5 °C to 40 °C (s)	Required: ± (0.6 + 0.01 $\| t_{wb}$)		Response time in air
	Desirable: ± (0.05 + 0.002$\|t_{wb}\|$	≤1	

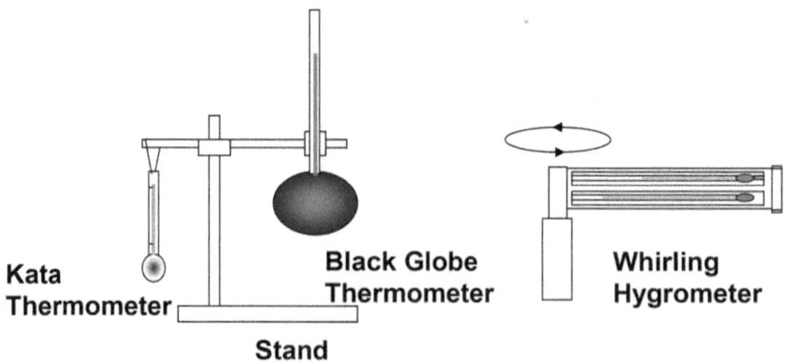

FIGURE 4.2. The set of instruments in the 'Admiralty box'. A kata thermometer provides cooling time; a black globe thermometer (globe temperature); and a whirling hygrometer containing a dry bulb thermometer (dry bulb temperature) and a wet bulb thermometer (wet bulb temperature). Charts are used with measurements to derive air temperature, mean radiant temperature, air velocity and humidity.

CLOTHING

Most people wear clothing most of the time. It reduces heat transfer between the body and the environment. If a person wears too much clothing they will become hot and too little clothing, cold. In any thermal environment, there is a required amount of clothing that ensures survival and a more tightly specified one that ensures comfort.

Heat exchange between the body and the environment through clothing is a dynamic property that can vary with body posture and movement as well as with active and smart control systems, the adjustment of clothing ensembles and garments, as part of adaptive opportunity, and more.

Clothing is usually initially specified in terms of the dry thermal insulation it provides and is usually estimated from a table of similar garments and clothing ensembles that have been measured on a heated thermal manikin (Table 4.5 and ISO

TABLE 4.5
Individual clothing garments: dry thermal insulation values

Garment description	Thermal insulation Clo (I_{clu})
Underwear	
Panties	0.03
Underpants with long legs	0.10
Singlet	0.04
T-shirt	0.09
Shirt with long sleeves	0.12
Panties and bra	0.03
Shirts/blouses	
Short sleeves	0.15
Lightweight, long sleeves	0.20
Normal, long sleeves	0.25
Flannel shirt, long sleeves	0.30
Lightweight blouse, long sleeves	0.15
Trousers	
Shorts	0.06
Lightweight	0.20
Normal	0.25
Flannel	0.28
Dresses/skirts	
Light skirt (summer)	0.15
Heavy dress (winter)	0.25
Light dress, short sleeves	0.20
Winter dress, long sleeves	0.40
Boiler suit	0.55
Sweaters	
Sleeveless vest	0.12
Thin sweater	0.20
Sweater	0.28
Thick sweater	0.35

(Continued)

TABLE 4.5
(Continued)

Garment description	Thermal insulation Clo (I_{clu})
Jackets	
Light summer jacket	0.25
Jacket	0.35
Smock	0.30
High insulative, fibre-pelt	
Boiler suit	0.90
Trousers	0.35
Jacket	0.40
Vest	0.20
Outdoor clothing	
Coat	0.60
Down jacket	0.55
Parka	0.70
Fibre-pelt overalls	0.55
Sundries	
Socks	0.02
Thick ankle socks	0.05
Thick long socks	0.10
Nylon stockings	0.03
Shoes (thin soled)	0.02
Shoes (thick soled)	0.04
Boots	0.10
Gloves	0.05

From: Olesen and Dukes-Dubos (1988). Example office ensemble: shoes + socks + underwear + trousers + shirt + tie + jacket = 0.03 + 0.02 + 0.09 + 0.25 + 0.25 + 0.01 + 0.35 = 1.0 clo

9920, 2007). In an environmental ergonomics survey, details of a uniform or the clothing garments and ensembles worn by people as well as clothing behaviour (adding, removing or adjusting) are required to provide an estimate of their thermal properties and opportunities.

Goldman (2006) considers the four Fs of clothing as fashion, feel, fit and function. The influence of fashion, in particular, should not be underestimated. People will often readily sacrifice thermal comfort for fashion and unacceptable fashions, particularly related to a uniform, may dominate feelings of dissatisfaction.

A simple model of clothing is to consider it as a layer of dry insulation (intrinsic insulation, Icl) covering the body, surrounded by an air layer that provides insulation (Ia) and is at a temperature from the clothing surface temperature to the environmental temperature. The air layer is affected by environmental conditions and the intrinsic layer is not. The intrinsic clothing insulation can be estimated by comparing the clothing on people in the environmental ergonomics survey with ensembles or garments that have been derived from measurements made on a heated thermal manikin

(Table 4.5). The insulation of the air layer is usually estimated from heat transfer by convection (air temperature and velocity).

Clothing insulation is measured as a thermal resistance per square metre of the body surface area with units m^2 °C W^{-1}. Gagge et al. (1941) first proposed the clo unit to provide a practical indication of clothing required by soldiers. 1 clo was the clothing insulation required for comfort in an office-type environment. It was said to be the insulation of a typical business suit and given the value of 1 clo = 0.155 m^2 °C W^{-1}. Zero clo indicated a nude person and 2 clo and above could be regarded as winter clothing.

For more specialist environmental surveys, for example, in hot conditions where sweating may occur, or where impermeable clothing is worn, then the vapour permeability of clothing will be important. Parsons (2014) considers dry thermal insulation; vapour and moisture permeability; compressibility and pumping and ventilation properties as well as fit, colour and behavioural affects such as the ability to add, remove and adjust clothing, change posture and more. ISO 9920 (2007) provides tables and methods for estimating the thermal properties of clothing and Goldman and Kampman (2007) and Parsons (2014) provide details.

METABOLIC HEAT PRODUCTION

If we burn 1 gram of carbohydrate in oxygen, it will be converted to carbon dioxide and water and produce on average around 4 kcal (circa 17 kJ) of heat. One gram of protein produces 4 kcal (17 kJ) and one gram of fat, 9 kcal (37 kJ) (the 4:4:9 method used as an approximation in food labelling). For people, it is assumed that, in each living cell, the food we eat is burned in the oxygen we breathe and the rate at which energy is produced adds up, across all cells, to the metabolic rate.

The relative amount of energy produced in tissues depends upon the activity of the person. Muscles produce relatively large amounts of energy during physical activity, and the blood supply distributes heat around the body to provide a representative metabolic heat production for the person.

Food is typically made up of carbohydrate, protein and fat (with alcohol at 7 kcal per gram). The ratio of the CO_2 produced to O_2 consumed is a measure of combustion with values of 1, 0.8 and 0.7 for carbohydrate, protein and fat, respectively, and an estimate of 0.85 for a 'normal' diet.

McIntyre (1980) suggests that for a 'normal' mixed diet, the heat equivalent of one litre of oxygen is 20.6 kJ (4.9 kcal). The heat production can then be estimated from breathing rate and the difference between the oxygen in inspired (20.93%) and expired air.

For the environmental ergonomics assessment of a thermal environment, it is useful, and for some methods essential, to estimate the metabolic heat production of the people who occupy the environment (Table 4.6). ISO 8996 (2004) provides methods of estimation. To allow comparison between individual people and groups, the estimate of metabolic heat production is 'normalised' by dividing the value by an estimate of the body surface area (often taken as 1.8 m^2 (around 2.0 to 1.6 m^2 from large to small adults and populations)).

One met is a unit based upon the metabolic rate of a standard resting person and is given the value of 50 kcal m^{-2} hr^{-1} later converted to 58.15 Wm^{-2}. Basal metabolic

TABLE 4.6
Estimates of metabolic rate for basic activities

Basic activity	Estimate of metabolic rate (W m^{-2})
Lying	45
Sitting	58
Standing	65
Computer operation	70
Driving	70–100
Walking on level even path at 2 km h^{-1}	110
Walking on level even path at 5 km h^{-1}	110
Going upstairs (0.172m/step), 80 stairs per minute	440
Transporting a 10 kg load on the level at 4 km h^{-1}	185

rate is the realistically lowest value for a fasting person lying down and is around 45 Wm^{-2}. Values for work are provided by Parsons (2014) as: 65; 100; 165; 230; and 290 Wm^{-2} for resting, low, moderate, high and very high levels of activity, respectively. Interpolation provides a method of estimation, usually sufficiently accurate, for the environmental ergonomist.

Metabolic heat production has a great influence on the human response to any thermal environment. The preservation, distribution and disbursement of the metabolic heat will determine the effects any thermal environment will have on a person. Insufficient heat loss will cause heat strain, too much heat loss will cause cold strain and just the right amount will preserve thermal comfort. An analytical approach to determining which of the above states is achieved and how it can be achieved is provided in the thermal audit.

THE THERMAL AUDIT

The thermal audit quantifies heat inputs to the body and heat outputs from the body. In summation, it can then determine heat storage. If the storage is positive, then there is a tendency for the body temperature to rise; if it is negative, to fall; and if it is zero, there may be thermal strain but homeostasis will be achieved (Figures 4.3 and 4.4).

Scenario testing (usually on a computer) can manipulate heat inputs, heat transfer and heat outputs to provide useful estimates such as when heat stress or cold stress limits will be exceeded, how much clothing is needed for survival and comfort, when dehydration will occur and what values of environmental variables (e.g. air temperature) are optimum and more.

The thermal audit begins with a conceptual body heat equation starting with metabolic heat production (H) and heat loss from the body to the environment by conduction (K), convection (C), radiation (R); and evaporation (E), the summation providing heat storage (S). By convention, heat production is always positive, heat gain to the body by heat transfer is negative (heat loss is positive) and heat storage is therefore

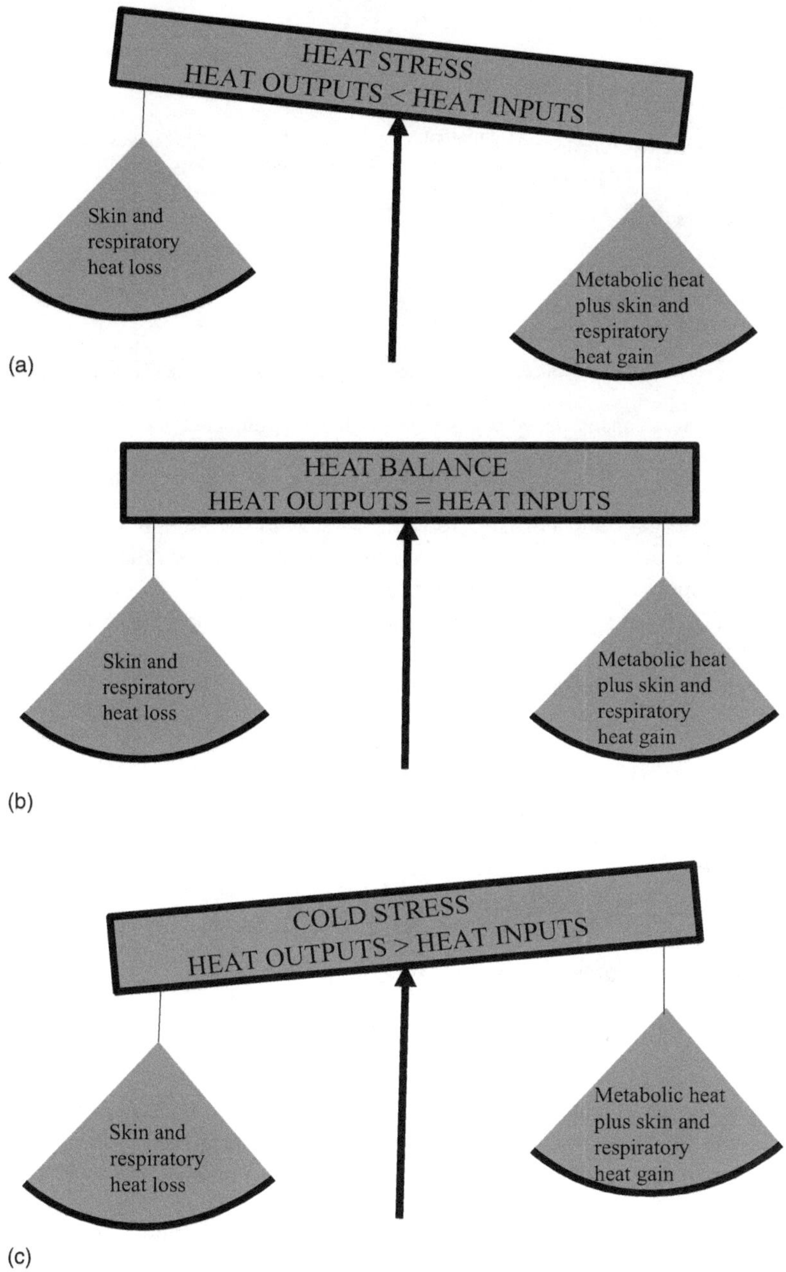

FIGURE 4.3 (a–c) A heat balance showing heat stress where the sum of heat gains (inputs) is greater than the sum of heat losses (outputs) (Figure 4.3a). Heat balance, where the sum of heat gains is equal to the sum of heat losses (Figure 4.3b); and cold stress where the sum of heat gains is less than the sum of heat losses (Figure 4.3c).

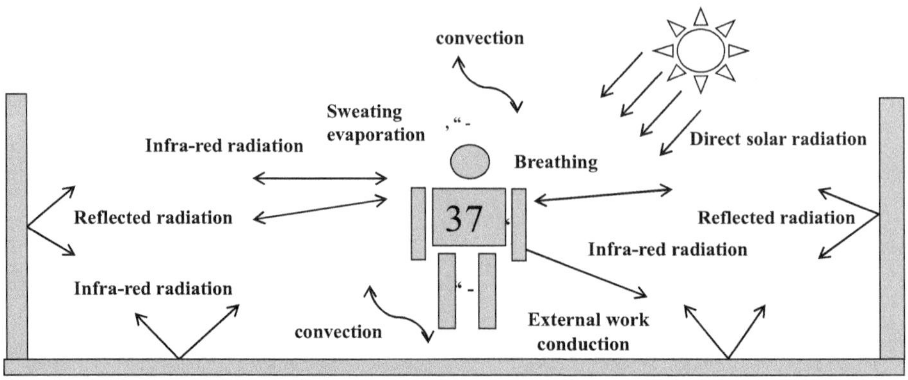

FIGURE 4.4 Heat transfer between a person and the environment.

positive for heat gain and negative for heat loss. All components are calculated in Watts per square metre of the body surface area (Wm^{-2}). To calculate body temperature rise, the specific heat of the body can be estimated as 3.49 kJ kg^{-1} K^{-1} and a sweat rate of 0.26 litres per hour (kg per hour) provides around 100 Wm^{-2} of heat loss by evaporation.

The conceptual equation for the thermal audit is

$$H - K - C - R - E = S$$

or for heat balance, to calculate required values for homeostasis and environmental design;

$$H - K - C - R - E = 0$$

The practical body heat equation for the thermal audit is

$$H - (C + R + Esk) - (Cres + Eres) = S \quad (4.3)$$

That is, metabolic heat production due to activity (H) minus heat loss at the skin (C +R + Esk) minus heat loss by breathing (Cres + Eres) provides heat storage. So, for example, for a person walking producing 100 Wm^{-2} of heat and heat loss at the skin adds up to 80 Wm^{-2} and heat loss by breathing is 5 Wm^{-2}, then this sums to a storage of 100 – 80 – 5 = 15 Wm^{-2}. So body temperature will tend to rise and thermoregulatory mechanisms such as sweating will be invoked to return S to 0 Wm^{-2} and homeostasis.

THE THERMAL AUDIT OF AN OFFICE WORKER

The environmental ergonomist measured the thermal environment in an office as 25 °C for air and mean radiant temperature ($t_a = t_r$); 50% relative humidity and

0.15 ms^{-1} air velocity. Workers wore light clothing and sat at desks operating computers. A thermal audit was conducted as follows:

$$\text{Metabolic heat}: H = M - W$$

It is clear that not all metabolic energy produced (M) by the body is heat as it requires mechanical energy to conduct work (W) as part of body movement and systems. This is usually a small part of energy production and for a resting person can be regarded as zero and less than 10% for most activities but up to around 20% for cycling. In general, H = M – W. For a person at work on a computer, we may estimate metabolic rate as M = 65 W m^{-2} so

$$H = M - 0 = 65\,Wm^{-2}.$$

$$\text{Conduction}: K = 0.$$

Most of a seated person is in contact with surfaces of high thermal insulation and low thermal conductivity. These include air as well as clothing and the surfaces of the chair and floor. So, it is reasonable to assume that heat transfer by conduction is negligible and

$$K = 0.$$

$$\text{Convection}: C = \text{fcl hc}\left(\text{tcl} - t_a\right)$$

Convection is heat transfer by the movement of air around the body driven by a difference in temperature between the clothing surface and the air. The hc value is the convective heat transfer coefficient and depends upon air velocity to force heat transfer. With no forced velocity, the air naturally removes heat as it becomes less dense at the clothing surface and rises. A value of hc = 3.1 W m^{-2} K^{-1} is an estimate for natural convection. The mean temperature of clothing is calculated using a computer in an iterative procedure but will be at air temperature for high air velocity and towards a mean skin temperature of tsk = 33 °C for light clothing and no forced air velocity, depending upon radiant temperature. For $t_a = tr = 25$ °C and an office environment an estimate of tcl = 28 °C seems reasonable. A Clo value assumes that the clothing is evenly distributed across the whole body. It will therefore increase the surface area for heat exchange by fcl = 1 + 0.3 clo. So for light clothing of 0.8 clo, fcl = 1 + 0.3 × 0.8 = 1.24. Heat transfer by convection can then be estimated as

$$C = 1.24 \times 3.1 \times \left(28 - 25\right) = 11.5\,W\,m^{-2}$$

$$\text{Radiation}: R = \text{fcl hr}\left(\text{tcl} - t_r\right)$$

Heat transfer is exchanged between the body and the surfaces of the environment driven by the difference in the fourth power of the absolute temperatures of the clothing temperature and the mean radiant temperature of the surroundings. The linear approximation to this provides an hr value related to the emissivity of clothing, the Stefan-Bolzman constant and the projected area of the body with respect to any radiation sources. For an office environment, hr is often taken as 4.7 W m^{-2} K^{-1}. Heat loss by radiation from the body is then

$$R = 1.24 \times 4.7 \times (28 - 25) = 17.5\ \text{W m}^{-2}$$

$$\text{Evaporation}: E = w(\text{Psk,s} - \text{Pa}) / (\text{Re,cl} + 1 / \text{fcl.he})$$

The driving force for evaporation of sweat from the body can be represented by the partial vapour pressure difference between the saturated vapour pressure at skin temperature (Psk.s = 5.0 kPa at a skin temperature of 33 °C) and the partial vapour pressure in the air (Pa = 1.6 kPa at a temperature of 25 °C and 50% relative humidity). For a nude person who is completely wet with sweat, the maximum evaporation will occur. Emax = he (Psk, s – Pa). The Lewis relation is he/hc = 16.5 K kPa^{-1}. So, he = 16.5 × 3.1 = 51.1 Wm^{-2} kPa^{-1} and Emax is 51.1 × (5.0 – 1.6) = 173.74 Wm^{-2} for a nude person in this environment. Evaporation of 0.26 litres of sweat per hour will release 100 Wm^{-2}.

For a clothed person, the vapour permeability of the clothing and the air layer around the body must be taken into account. For typical clothing, this is around 0.015 m^2 kpa W^{-1} and for air 1/fcl.he = 1/(1.24 × 51.1) = 0.016 so a total vapour permeability resistance of 0.031 m^2 kPa W^{-1}. So Emax = (Psk,s – Pa)/(Re,cl + 1/fcl.he) = 3.4/0.031 = 109.7 W m^{-2}. For an office environment and thermal comfort, we can assume no active sweating and that the evaporation from the body, divided by the maximum evaporation that can occur (skin wettedness), w = 0.06. So

$$E = 0.06 \times (5.0 - 1.6) / (0.015 + 0.016) = 6.6\ \text{W m}^{-2}$$

$$\text{Breathing}: \text{RES} = \text{Cres} + \text{Eres}.$$

When we breathe, the inspired air is heated from air temperature to body temperature and it is saturated in the lungs at internal body temperature when it was previously at the humidity and temperature of the air. The 'conditioned' air is then expired, so breathing is a mechanism for heat exchange by convection (Cres) and evaporation (Eres). For an office environment, we can assume that the total heat loss by breathing, from a person, is around 5 W m^{-2}.

$$\text{RES} = 5\ \text{Wm}^{-2}$$

So the total thermal audit is

$$H - K - C - R - E - \text{RES} = S$$

$$65 - 0 - 11.5 - 17.5 - 6.6 - 5 = 24.4\ \text{W m}^{-2}$$

Thus, the thermal audit indicates that a person will respond to a heat gain of 24.4 Wm^{-2}.

INTERPRETATION

The simple thermal audit outlined above implies that a person in the office would need to increase heat loss to maintain thermal balance. It is probable that this could be achieved through increased heat loss by convection and radiation. The mechanism would be vasodilation with increased blood flow to the skin to increase skin temperature and create a sensation of warmth with some dissatisfaction. To reduce dissatisfaction, mechanisms for increasing air velocity, decreasing air temperature and allowing adjustments to clothing should be explored within the context of the environmental ergonomics investigation.

The power of the thermal audit is that it provides a method for understanding the mechanisms of human response to any thermal environment, allowing interpretation and interaction to make recommendations. The environmental ergonomist 'gets a feel for' the thermal environment and probable human response to it. A computer programme can provide a more detailed thermal audit with more detailed and flexible interaction (Parsons, 2014).

The thermal audit provides the basis for internationally accepted analytical methods, used in guidelines and regulations. ISO 7933 (2004) calculates the sweating required in a hot environment as well as allowable exposure times and more. ISO 7730 (2024) provides a prediction of thermal sensation based upon thermal load and heat storage and ISO 11079 (2007) uses a heat balance equation to determine how much clothing insulation is required in cold conditions. Combined with a representation of the properties of the human body, and a controlling thermoregulation system, the thermal audit provides an integral part of computer models for assessing human response to the thermal environment. The above methods are used in the case studies presented in the following chapters.

FURTHER READING

Parsons K C 2014; *Human thermal environments* 3rd Edition), CRC Press, Taylor & Francis Group, Boc Ratan, Oxford, New York ISBN 978-1-4665-9599-6

McIntyre D. A., 1980, *Indoor climate, Applied Science*, London, ISBN 0-85334-868-5.

STANDARDS

ISO 7726, 1998; *Thermal environments – Instruments and methods for measuring physical quantities*, ISO, Geneva

ISO 7730, 2005; *Ergonomics of the thermal environment – Analytical determination and interpretation of thermal comfort using calculation of the PMV and PPD indices and local thermal comfort criteria*, ISO, Geneva

ISO 7933, 2004; *Ergonomics of the thermal environment – Analytical determination and interpretation of heat stress using calculation of the Predicted Heat Strain*, ISO, Geneva.

5 Case Study
Thermal Comfort in an Open Plan Office

ENQUIRY

Fred is a manager of a retail management company with an open plan accounting office covering one floor of a city high-rise building. He contacted Sheila who is an environmental ergonomist at Oxbridge University. Workers in the office were complaining of thermal discomfort and suggesting that it was affecting their work. They were threatening to cease work and there had been some absenteeism as well as requests to work from home.

It was important that the problem is rectified and that thermal comfort for all workers can be achieved. An independent expert investigation would demonstrate to workers that the problem is being taken seriously by the management as it is important that something is done and is seen to be done.

THE MEETING

By exchange of email and a video conversation, Sheila suggested a meeting with Fred in the building where the office is located at a time when problems may occur. Fred suggested that to reduce costs and travelling time, the meeting should take place as a briefing immediately before the thermal comfort survey is carried out. A standard thermal comfort survey was required with a possible follow-up if necessary. A proposal with costs was prepared and agreed for an initial one-day survey.

Sheila and her team (Paul and Lisa) with (calibrated) equipment attended a meeting in the morning at the start of the working day. They were met by Fred, Alice, the workers' representative and William, the building services engineer in charge of the building's heating, ventilation and air conditioning systems.

At the meeting, Fred explained that up to 50 computer workplaces are in partition cubicles and on long back-to-back desks with sealed windows to one side. Workers are complaining of thermal discomfort and headaches, particularly in the afternoon. William explained that he was surprised at the complaints as a sophisticated displacement ventilation and comfort control system had been installed that floated cool air along the floor through vents that rose to the ceiling as it warmed, taking any pollutants and infections with it. Alice explained that this problem should not be taken lightly as it was causing serious distraction and worker dissatisfaction.

Sheila explained and demonstrated the procedures for the thermal comfort survey. Measurements of air temperature and humidity will be taken using a device like an old-fashioned football supporter's rattle (whirling hygrometer); Radiant temperature

DOI: 10.1201/9781003401964-7

will be assessed with a black globe that looks like an old-fashioned toilet ball cock, with a temperature sensor inside it, and a children's bubble machine, along with an electronic device, will be used to assess and measure air movement (see Figure 4.2).

Paul will use those instruments and will attempt to measure the environment in all workplaces or use a sampling or grid system to obtain measures representative of all workstations. Lisa is responsible for determining the thermal comfort responses of workers that will mainly involve subjective responses in a brief questionnaire including how workers feel now and how they generally feel at work with a comment box for open-ended responses. All workers present in the office will be asked to complete the questionnaire and their location will be noted.

Sheila will require the previously requested blueprint plan of the office with as much detail as possible, particularly about thermal sensors and control systems, as well as dimensions and workplace positions. Measurements will also be made independently of the dimensions and layout of the office. Sheila will also complete an expert observation checklist questionnaire, noting anything relevant to the thermal comfort of the workers and their opportunities to achieve thermal comfort if they become uncomfortable and if they were taking them. She will also note the clothing worn by the workers and their type of physical activity (computer work, sitting, standing, walking, etc.). Fred noted that the clothing dress code was smart casual and of the 30 workers in the office that day, 20 identified as female and 10 as male (standard underwear was assumed).

All equipment and materials were shown to Alice who agreed that the procedures were acceptable and would be explained to the workers. Alice agreed to explain to the workers that the survey was to improve their thermal comfort, that it was their individual responses that were required and that asking the opinion of others and group discussions were not useful.

Two complete surveys would be completed taking around two hours each. One in the morning and one in the afternoon. Neither Fred nor William would be present during the survey and a helpful disposition will be taken by all involved. William would ensure that the heating, ventilation and air conditioning systems were set as normal.

It was agreed that all should meet after the survey was complete to provide comments and additional information if required. A report marked 'commercial in confidence', would be sent to Fred by an agreed date.

Agreed Aims and Objectives

Aims and objectives were agreed before the meeting in a written proposal and they were confirmed at the meeting as follows

Aim

To conduct a thermal comfort survey and assessment of the open plan office and to provide recommendations for ensuring the thermal comfort of workers in the future

Objectives

Objective 1. To quantify the thermal environment and conditions that influence thermal comfort in the office

Objective 2. To determine the existing thermal comfort and discomfort of office workers and their responses by both quantitative and qualitative methods
Objective 3. To specify thermal comfort conditions for the office based upon accepted national and international standards for thermal comfort
Objective 4. To compare the conditions in the office with those specified in accepted thermal comfort standards
Objective 5. To identify the causes of any thermal discomfort and make proposals to provide thermal comfort
Objective 6. To make overall recommendations to ensure thermal comfort in the future, including any further investigations and actions that need to be taken

THE ENVIRONMENTAL ERGONOMICS THERMAL COMFORT SURVEY

Sheila used the blueprint of the office to determine where measurements should be taken that would be representative of the thermal environment experienced by workers. Ten locations were identified for physical measurement and all workers in occupied workplaces would be provided with a simple one-page questionnaire to complete at their desks with Lisa present. Sheila would make observations using a checklist that included worker behaviour as well as adaptive opportunities and gain an overall impression of the environment, including the perceived social environment as well as overall quality.

METHODS TO ACHIEVE OBJECTIVES

Objective 1. To quantify the thermal environment and conditions that influence thermal comfort in the office

Paul placed his five black globe thermometers on the desk at five of the previously chosen workstations noting initial readings and being careful to ensure that they would be influenced by any thermal radiation that would affect a person at the workplace. A black globe thermometer indicated whether the worker is influenced by thermal radiation (from 'hot' computers, the sun, cold windows, etc.) and provides a measure of mean radiant temperature when globe temperature is 'corrected' for air temperature and air velocity (Parsons, 2014). The black globe thermometer (150 mm diameter) requires around 20 minutes to 'settle' before a valid reading is taken.

After the first five globe thermometers had been positioned, Paul used the whirling hygrometer to measure dry bulb temperature and wet bulb temperature at the workplace and a hot wire anemometer, bubbles and a simple temperature and humidity electronic instrument at ankle, chest (providing some cross-checking) and head height. Paul noted activity and clothing at each workstation while making the measurements.

After five workstations had been measured and 'final' globe temperatures were taken, the five globes were moved to the second set of workstations and measurements at those workstations taken. The morning procedure was repeated for the second set of measurements made in the afternoon. The first set of measurements was taken between 10.0 am and noon and the second set of measurements was taken between 2.0 pm and 4.0 pm.

Objective 2. To determine the existing thermal comfort and discomfort of office workers and their responses by both quantitative and qualitative methods

Lisa gave out a single-sheet thermal comfort questionnaire (see Figure 5.1) to all workers in the office with instructions for workers to complete it when she was present and while Paul was making the measurements. The questionnaire was described to each worker explaining that it was their own individual feelings that were required, that NOW meant at that exact time and GENERALLY meant an overall assessment of their feelings at that time in the office not just on that day but overall. The person was identified by name, initials or location depending upon their preference and individual information and responses would be kept confidential to Lisa and not included in any report. Clothing and activity were also noted by Lisa for cross-checking. Two surveys (Paul and Lisa) were completed in parallel with the measurements described for objective 1 above.

Lisa also carried out an observation checklist of the office (Figure 5.2).

Objective 3. To specify thermal comfort conditions for the office based upon accepted national and international standards for thermal comfort

Please let us know your thermal comfort in the office.

Please indicate how YOU feel NOW.

- Hot
- Warm
- Slightly warm
- Neutral
- Slightly cool
- Cool
- cold

Please indicate how YOU would like to be NOW in the office

Warmer No change Cooler

Indicate why you gave those responses or any other Information about YOUR thermal comfort (draughts, dryness, stickiness..)

Please indicate how you GENERALLY feel in the office

- Hot
- Warm
- Slightly warm
- Neutral
- Slightly cool
- Cool
- cold

Please indicate how YOU would generally like to be in the office

Warmer No change Cooler

Indicate why you gave those responses or any other Information about YOUR thermal comfort (draughts, dryness, stickiness…)

Are you generally satisfied with your thermal environment in the office? Yes No

FIGURE 5.1 One-page thermal comfort questionnaire used in the office.(see also extended version described for objective 2 below)

Observation checklist for the thermal environment. To be completed by the investigator in the office.

General impression of the thermal environment: One sentence and one word descriptors.

What is best about the thermal environment in the office ?

What is worst about the thermal environment in the office ?

Would you say that the office was hot, cold or neutral and comfortable for the workers?

Looking around the room would you say that there are better areas and worse areas for thermal comfort ?

Where would you prefer to be located?

Make a note of the activities people are carrying out in the office for each workplace.

Make a note of the clothing worn by each person in the office and note their location

Note personal adaptive opportunities for people to achieve thermal comfort

	Opportunity	Opportunity taken
Adjust clothing		
Move around		
Move out of the office		
Change workplace		
others		

Note technical adaptive opportunities for people to achieve thermal comfort

	Opportunity	Opportunity taken
Adjust thermostat		
Adjust workplace conditions		
Open or close windows		
Adjust blinds or curtains		
others		

Does thermal environment cause discussion or distraction?

Overall conclusion: Is thermal environment optimum ? Yes/No Recommendations at this stage……

FIGURE 5.2 Observation checklist used by the investigator (Lisa) in the office, while people were working.

Two thermal comfort standards are prominent worldwide; ASHRAE standard 55 and ISO 7730 (2024). They provide similar methods and results. ISO 7730 (2024) is based upon the work of Fanger (1970) who conducted work in Kansas, USA, and Copenhagen, Denmark. He integrated his own research with that of others to develop a thermal comfort index on the assumption that steady-state thermal comfort is determined by a person being in heat balance but also that skin temperatures and sweat rates must be within comfort limits.

Based upon those premises, he formulated a comfort equation that specified conditions for comfort and using the concept of thermal load (related to the difference between comfort conditions and actual conditions), developed it into a practical index that predicted the mean rating a large group of people would give in 'those conditions' if presented with the rating scale for sensation: +3, hot; +2, warm; +1, slightly warm; 0, neutral; −1, slightly cool; −2, cool; −3, cold.

The value was termed the predicted mean vote (PMV) and it was derived from values of air temperature, mean radiant temperature, air velocity and humidity as well as clothing insulation and metabolic heat production. A predicted percentage dissatisfied (PPD) value was derived from large-scale laboratory data where the predicted percentage of people outside of the range PMV = ±2 were assumed to be dissatisfied. Examples of PMV values are provided in Table 5.1.

TABLE 5.1
Predicted mean vote (PMV) values. Assume working at a desk, rh = 50%; still air; and $t_a = t_r$. PMV: +3, hot; +2, warm; +1, slightly warm; 0, neutral; –1, slightly cool; –2, cool; –3, cold (Fanger, 1970)

Icl, Clo							
$t_a = t_r$	**0.1**	**0.3**	**0.5**	**0.8**	**1.0**	**1.5**	**2.0**
10					–2.7	–1.6	–0.9
12				–2.8	–2.2	–1.2	–0.6
14				–2.3	–1.8	–0.9	–0.3
16			–2.8	–1.8	–1.3	–0.5	0.0
18		–2.9	–2.1	–1.2	–0.8	–0.1	0.3
20		–2.2	–1.5	–0.7	–0.4	0.2	0.6
22	–2.3	–1.4	–0.8	–0.2	0.1	0.6	0.9
24	–1.4	–0.7	–0.2	0.3	0.6	1.0	1.3
26	–0.5	0.1	0.4	0.8	1.0	1.4	1.6
28	0.4	0.8	1.1	1.3	1.5	1.7	1.9
30	1.3	1.5	1.7	1.8	1.9	2.1	2.2
32	2.0	2.1	2.2	2.3	2.3	2.4	2.4

ISO 7730 (2024) also provides details and equations for predicting the percentage dissatisfied due to 'local thermal discomfort' caused by draughts, for example. Conditions that provide a PMV = 0 (neutral), as well as the absence of local thermal discomfort, will provide a specification of thermal comfort conditions for the office.

Objective 4. To compare the conditions in the office with those specified in accepted thermal comfort standards

The actual PMV and predicted local discomfort, based upon measured office conditions, were used to compare actual conditions with accepted thermal comfort standards.

Objective 5. To identify the causes of any thermal discomfort and make proposals to provide thermal comfort

Causes of thermal discomfort were determined by examining influences on air temperature; mean radiant temperature, air velocity, humidity, clothing and activity as well as limited opportunities for adaptive behaviour. Air movement at the feet was causing draughts.

Objective 6. To make overall recommendations to ensure thermal comfort in the future, including any further investigations and actions that need to be taken

Causes of thermal discomfort were considered to allow adjustment and opportunity for achieving thermal comfort.

The final meeting was held at 5.0 pm, after the survey had been completed. It was attended by Fred, Alice and Sheila. All were satisfied with the survey and looked forward to the report.

THE REPORT

Executive Summary

The aim of the project was to conduct a thermal comfort survey and assessment of the open plan office and to provide recommendations for ensuring the thermal comfort of workers in the future. Both physical measurements of the environment and subjective measures were taken, as well as general observations.

The environmental ergonomics thermal comfort survey found that air and radiant temperatures were around 18 °C in the morning and 22 °C in the afternoon. Relative humidity was around 50% and there was still air in both morning and afternoon. International standard predictions were on average slightly cool to cool in the morning and neutral to slightly cool in the afternoon with females cooler than males due to lighter clothing. Dissatisfaction due to draughts was predicted around the ankles in some workstations. Actual responses confirmed predictions. Around 90% were dissatisfied in the morning due to being too cold and 95% found the afternoon environment unacceptable due to headaches.

Recommendations included a re-design of the displacement ventilations system, a raising of the air temperature to 23 °C on average with a recommendation that if workers were too cool then they should increase their clothing level. Opportunities for allowing workers to achieve thermal comfort (e.g. more opportunity to move around) should be explored.

MEETING OBJECTIVES

Objective 1. To quantify the thermal environment and conditions that influence thermal comfort in the office

The thermal environment was listed at each workstation on a plan of the office in terms of air temperature, mean radiant temperature, air velocity and humidity. Air temperature was the dry bulb temperature of the whirling hygrometer and together with aspirated wet bulb temperature humidity was derived from a psychrometric chart (Figure 4.2 and 4.4). As indicated on the plan, for most of the workstations, globe temperature and hence mean radiant temperature was similar to air temperature indicating no significant thermal radiation at that workplace. Relative humidity was consistently around 50% and most workstations had no perceptible air movement, traditionally given a value of 0.15 ms^{-1}.

Activity ranged from around 60 Wm^{-2} for sitting at rest to 70 Wm^{-2} for computer work providing a realistic average of 65 Wm^{-2} for all of the workers in the office. Clothing ranged from estimates of 1.0 Clo for males and 0.75 Clo for females. Workstations near south-facing windows received solar radiation in the afternoon but workers generally did not adjust blinds. The outside air temperature was 20 °C in the morning and 25 °C in the afternoon.

Sheila noted that at some desks air was supplied through vents at 0.3 ms^{-1} at ankle height and at a temperature of 13 °C. Most workers, however, had covered the vents with files or paper to restrict the air flow. Males tended to take off their jackets in the afternoon but females tended not to adjust their level of clothing.

It could be seen from the plan that the average air temperature at chest height was around 18 °C in the morning and 22 °C in the afternoon.

The PMV and PPD values (ISO 7730, 2024) were calculated for all workstations and shown on a plan of the office for both morning and afternoon.

For average conditions in the morning, at 18 °C air temperature (ta), 18 °C mean radiant temperature (t_r); 0.15 ms^{-1} air velocity (v); and 50% relative humidity (rh) (1031 Pa partial vapour pressure (pa)) and an activity level of 65 Wm^{-2}, for 0.75 clo of clothing insulation, the PMV = −1.8 (cool) and the predicted percentage of dissatisfied, PPD = 68%. For 1.0 clo, PMV = −1.2 (slightly cool) and the PPD = 35%.

For average conditions in the afternoon, at 22 °C air temperature (t_a), 22 °C mean radiant temperature (t_r); 0.15 ms^{-1} air velocity (v); and 50% relative humidity (rh) (1322 Pa partial vapour pressure (pa)) and an activity level of 65 Wm^{-2}, for 0.75 clo of clothing insulation, the PMV = −0.7 (slightly cool) and the PPD = 15%. For 1.0, clo PMV = −0.2 (neutral) and the PPD = 6%.

The Draught Rating (%of people in whole-body comfort but bothered by draught) for an air velocity of 0.3 ms^{-1} and turbulence intensity of 40%, is 76%, for people in 18 °C air temperature (probably higher than that due to their cooler disposition) and 57% for people in 22 °C air temperature (see ISO 7730, 2024). For 13 °C, air temperature at floor level a dissatisfaction level towards 100% would be expected for exposed skin.

Objective 2. To determine the existing thermal comfort and discomfort of office workers and their responses by both quantitative and qualitative methods

Lisa supervised the administration of the questionnaire at the time that the physical measurements were taken but did not intrude on the process. The instruction was to please make a mark on the scale how YOU feel. The scales related to how YOU feel in the thermal environment NOW and repeated for GENERALLY AT THIS TIME AT WORK for sensation (+3, hot; +2, warm; +1, slightly warm; 0, neutral; −1, slightly cool; −2, cool; −3, cold); preference (warmer, no change, cooler); uncomfortable (0, not uncomfortable; 1, slightly uncomfortable; 2, uncomfortable; and 3 very uncomfortable) and the same four-point scale structure for dryness; stickiness and draughty. After each scale was an open-ended comments space for additional information (asking to note ratings for areas of the body if relevant, most effective measure that could improve their thermal comfort and any other comments). Forced bipolar scales of satisfied/dissatisfied and acceptable/unacceptable concluded the single sheet questionnaire (ISO 28002, 2012).

Sheila noted that most workers remained at their workstations between breaks. Some males removed jackets in the afternoon. Blinds on the windows were not adjusted and some workers had covered the intake ventilation vents in the floor with bags, paper, etc. and when asked why, they reported that it was to reduce draughts.

The questionnaire results were summarised on a series of office plans with ratings corresponding to responses at workstations for morning and afternoon. The mean ratings indicated that workers were cold in the morning and cool in the afternoon. Some were neutral and not uncomfortable but no one was warm. They were very uncomfortable in the morning and uncomfortable in the afternoon. They preferred to be warmer and were slightly dry but not sticky.

Most complained of draughts, particularly in the morning, and those that complained, felt cold around feet and ankles. Some workers suggested that decreasing the rate of air flowing across the floor and increasing temperature would be most important for achieving thermal comfort and that they would be happy to work at home if this were not possible.

The general view was that the office on that day generally reflected their thermal (dis)comfort but some said that it was worse in the winter particularly near to windows; 90% of workers were dissatisfied in the morning with 95% rating 'unacceptable'; 70% were dissatisfied in the afternoon with 95% rating unacceptable. Headaches were reported particularly in the afternoon with difficulty in ability to concentrate.

Objective 3. To specify thermal comfort conditions for the office based upon accepted national and international standards for thermal comfort

Accepted standards for comfort across the world are given in ISO 7730 (2024). For conditions in the office where air temperature is equal to mean radiant temperature, there is still air and relative humidity is 50%, for an average metabolic rate of 65 Wm^{-2} and male and female clothing of 0.75 clo, the air temperature for whole-body comfort is 24.5 °C. For subjects wearing 1.0 clo (all male) the air temperature for comfort is 23 °C.

The main cause of local discomfort in the office is due to draught. ISO 7730 (2024) provides a minimum draught rate (predicted dissatisfied) based upon a turbulence intensity of 40% at an air temperature of 24.5 °C and at 23 °C if the air velocity is less than 0.2 ms^{-1}.

Objective 4. To compare the conditions in the office with those specified in accepted thermal comfort standards

In summary, the physical measurements of the thermal environment in the survey predicted that the office was too cool and draughty when compared with the comfort conditions specified in ISO 7730 (2024). In particular, the air temperature was between 18 and 22 °C in the office when international standards for thermal comfort indicated 24.5 °C.

Objective 5. To identify the causes of any thermal discomfort and make proposals to provide thermal comfort

A combination of all six basic parameters influences thermal comfort so a number of potential solutions exist to provide thermal comfort. These include increasing clothing insulation and activity levels, adding radiation and so on. Air temperature of 18 °C in the morning and 22 °C in the afternoon is too low for comfort and the obvious first step in achieving thermal comfort is to increase the air temperature to 24.5 °C. This will provide for whole-body comfort and allow some adaptability such as removing a jacket.

Air movement at floor level causes draughts and the current improvised solution of blocking vents is cumbersome and may be restricting the inflow of fresh air (allowing CO_2 levels to rise causing headaches and increased dissatisfaction in the afternoon). An engineering solution is required to prevent draughts at floor level, with almost still air on exposed skin around ankles.

Objective 6. To make overall recommendations to ensure thermal comfort in the future, including any further investigations and actions that need to be taken

Based upon the one-day environmental ergonomics survey conducted, we make the following recommendations:

1. Provide an engineering solution to ensure that the displacement ventilation system reduces air flow across the feet and ankles of the workers. A redesign of the office to move workplaces from vents or a redesign of the ventilation system should be considered.
2. Increase air temperature at chest height to 24.5 °C or to 23 °C if clothing can be increased to 1.0 clo for both males and females (see ISO 9920, 2007).
3. Give careful consideration to adaptive opportunities in the space to include a discussion with workers of clothing, curtains and blinds and redesign of the office.

Note: *Increasing air temperature may benefit the energy efficiency of the office as well as lowering the level of distraction that may be affecting productivity. It is also recommended that a measure of* CO_2 *level be made, particularly in the afternoon as this may be the cause of headaches and extra dissatisfaction.*

OUTCOMES

After the completion of the contract, the investigators were thanked for the report which was informative and met the aims and objectives. No feedback was provided on actions taken.

An internal review of the survey within the university concluded that the advantages of using online methods of subjective data collection should have been more clearly pursued. The internal managers did not wish to have computer work disturbed but a simple email system with a link to a questionnaire may have made the survey less intrusive and more efficient to carry out.

FURTHER READING

Parsons K C, 2020; *Human thermal comfort*, CRC Press, Boca Raton, ISBN 978-0-367-26193-1.

STANDARDS

ISO 7730, 2005; *Ergonomics of the thermal environment – Analytical determination and interpretation of thermal comfort using calculation of the PMV and PPD indices and local thermal comfort criteria*, ISO, Geneva

ISO 7726, 1998; *Thermal environments – Instruments and methods for measuring physical quantities*, ISO, Geneva.

ISO 8996, 2004; *Ergonomics of the thermal environment – determination of metabolic rate*, ISO, Geneva.

ISO 9920. 2007; *Estimation of thermal insulation and water vapour resistance of a clothing ensemble*, ISO, Geneva.

ISO 10551, 2019; *Ergonomics of the physical environment-Subjective judgement scales for assessing physical environments*, ISO, Geneva

ISO 28002, 2012, *Ergonomics of the physical environment-The assessment of environments by means of an environmental survey involving physical measurements of the environment and subjective responses of people*. ISO, Geneva.

6 Case Study
The Avoidance of Heat Casualties in Special Forces Selection Trials

ENQUIRY

David is an applied physiologist and manager with the Army. He has been contacted by a special forces recruitment officer (Sebastian) as they had experienced heat stress injuries and deaths during their selection process. David contacted Thomas (Professor of Environmental Ergonomics) at the Human Thermal Environments Laboratory at old Hartley University for advice.

THE MEETING

After an initial letter and exchange of emails between David and Thomas outlining the issue and requirements, a meeting was held at the special forces headquarters between Sebastian, David and Thomas. The meeting was to provide information and context and to determine specific aims and objectives for a study.

Sebastian explained that heat stress injuries and deaths had occurred during recruitment and that, in addition to the tragic outcome felt by all, the loss of valued personnel and colleagues and the waste of life, litigation had resulted in great expense, loss of reputation and a requirement to avoid heat casualties in the future.

Sebastian described the trial component of the current selection procedure. He emphasised that performing in extreme climates, including heat, was part of the requirement of special forces personnel. Avoidance of heat strain was not an option. Avoidance of chronic injury and death during selection was the requirement.

The selection process involved male and female soldiers in a two-day exercise to achieve objectives in mountainous terrain, often in extreme weather. The particular concern was for exercises on hot sunny days. Soldiers were very highly motivated, expected extreme risk and would not hesitate to place themselves in danger.

David explained that the scope of the project for the environmental ergonomics consultant was to provide advice so that his team of applied physiologists could develop a plan of action to support the trials. Details of the terrain, previous weather conditions and of the trial would be provided but access was not allowed. It was expected that the work of the environmental ergonomist would take a maximum of five days and that a report would be produced.

Thomas explained that international standards were available to indicate when unacceptable heat strain was likely and when it was unlikely. High physiological heat

DOI: 10.1201/9781003401964-8

strain, possible loss in cognitive performance, including confusion, as well as very high levels of motivation to complete the trial would require objective measures to be used. Behavioural measures may be useful but subjective measures used in isolation would be unreliable.

A report would be produced that outlined a plan that included procedures that should be adopted that would lead to the avoidance of unacceptable heat casualties. He would be available to advise on the implementation of the plan and he would expect the final procedures to have been established after user trials, with training for the support team whose job it was to ensure the avoidance of unacceptable heat strain.

The report providing advice may be regarded as stage 1 of the project. Stage 2 could involve training of staff involved in implementing the plan; identifying environmental conditions and monitoring the health and safety of the participants. Stage 2 may involve both laboratory and field tests and trials. Stage 3 could involve a series of simulated/actual trials to evaluate and develop the plan to its final acceptance. A database and full recording in a log of each actual trial will be useful for ongoing development.

AGREED AIM AND OBJECTIVES

The aim and objectives for stage 1 of the project were agreed in writing as follows:

AIM

To provide detailed advice on a procedure to ensure the avoidance of unacceptable heat casualties during field trials for the selection of special forces soldiers.

Objectives

1. To advise on instruments and methods for monitoring and measuring the heat stress of the environment as well as criteria for determining the risk of heat injury based upon heat stress
2. To advise on instruments and methods for monitoring and measuring the heat strain of participants subjected to the heat stress and on the criteria for determining the severity of heat strain including levels that may be considered unacceptable and a danger to health
3. To advise on methods that can be used to predict the likely heat strain of participants from the heat stress to which they are expected to be, and could be, exposed
4. To provide the advice in a report marked 'commercial in confidence', with an executive summary that states the aims and objectives, what was done to achieve those aims and objectives and conclusions related directly to aims and objectives

'A TEST SCENARIO'

Before the end of the meeting, Sebastian asked if Thomas would analyse a test scenario so that he could get a feel for the sort of predictive methods that would be used. It was hypothesised that a special forces officer set off on his own initiative in a hostile environment on mountainous terrain to travel from a beach location to a military base inland. It required a steep ascent up a cliff path for about 45 minutes, then along a fairly even path for an hour, circumnavigation around a hostile town for 30 minutes and then over a mountain to the base probably taking around 90 minutes. Conditions were hot starting at 35 °C air temperature rising to 48 °C by the end of the journey, all in full afternoon sun with no shade. The officer wore light military clothing and carried light ordinance as well as water for the journey and a small military non-reflective parasol.

The officer had been aware of the military adage 'don't be 'SHAFTED' by the heat' from the mnemonic 'SHAFTS' that is be Sensible, Hydrated, Acclimatised, Fit, Thin and Sober. All criteria were met apart from whether it was sensible to start on such a journey was not certain. The officer had full confidence in his ability to carry out the journey and was pleased to test himself. Was he foolish?

Thomas had his laptop computer with him and said that he could give an answer based upon three methods of analysis to come to a conclusion. A first consideration would suggest that it was not wise to undertake such a journey under normal circumstances and that for a highly motivated person, it was a dangerous enterprise. The three methods were to use the wet bulb globe temperature (WBGT) index to indicate the likelihood of damage to health (ISO 7243, 2017); an analytical index based upon the calculation of sweat rate required and its interpretation in the predicted heat strain (PHS) method (ISO 7933, 2023); and finally a computer model of human thermoregulation (Gagge et al., 1971). An estimate of the conditions was required and for the scenario, they were given as follows for each part of the journey (Table 6.1).

An indication of the consequences of undertaking the journey on the health of the officer can be estimated using a time-weighted average of the conditions over each

TABLE 6.1
Exposure time; environmental conditions; activity level (M); and clothing properties (Icl, Iecl, im) used to predict response to heat

			Environment				
Terrain	**time (min)**	**Duration**	**Metabolic rate**	**ta**	**tr**	**v**	**Pa**
Cliff ascent	0-45	45	400	35	60	0.5	
cliff path	45-105	60	150	38	60	1	
avoid town	105-135	30	100	40	65	0.8	
mountain ascent	135-195	60	400	45	55	0.8	
mountain decent	195-225	30	250	48	60	0.5	
Overall	0-225	225	273	41	59	0.75	0.77
rh	Icl	Iecl	im				
20	0.155	0.0155	0.38				

stage of the expedition. These are a metabolic rate of 273 Wm^{-2}, air temperature of 41 °C, mean radiant temperature of 59 °C and air velocity of 0.75 ms^{-1}. It is assumed that light hot weather uniform with ordinance is worn and conditions are dry with relative humidity low at 20% throughout. Some parts of the journey will be more stressful than others, for example, going uphill. It is assumed that no rests are taken (the power of predictive methods is that a number of scenarios can be tested so the following is a first approximation).

Predicting heat strain using ISO 7243 (2017) provides a natural wet bulb temperature of 24.5 °C and when integrated with 150 mm diameter, globe temperature of 56 °C and air temperature of 41 °C provides a WBGT value of 0.7 tnwb + 0.2 tg + 0.1 ta = 32.45 °C. For a weighted average metabolic rate of 273 Wm^{-2} that is well above the limit value for normal working conditions and for military exercises in the heat where a WBGT value of 25 °C would be a limiting value for an acclimatised man. There is clear danger of damage to health at 32.45 °C WBGT even for a person hydrated, acclimatised, fit, thin and sober. For avoidance of heat strain, it should be noted that high levels of motivation can lead to overconfidence and can be dangerous.

ISO 7933 (2023) provides a body heat equation method and was used to predict heat strain caused by the time-weighted average conditions. It was found that a limiting internal body temperature of 39 °C would be reached after only 26 minutes for a normal healthy male worker and that is probably a reasonable estimate for the SAS officer. Sweat loss and dehydration effects will occur within the time period of the journey even though water replacement was available.

A 25-node computer model of human thermoregulation (Stolwijk and Hardy, 1977; Parsons, 2021) was used to predict physiological response to the time-weighted average conditions every 10 minutes throughout the journey. Internal body temperature rose rapidly to above 39 °C after 120 minutes when the town was being avoided and at the start of the mountain ascent, and it is likely that the officer will start to become confused. After 210 minutes at the top of the mountain, it is likely that collapse will occur due to excessively high internal body temperature leading to death (see Table 6.2).

That would be the prediction. It is emphasised that the international standard methods are concerned with populations and not individual responses. The computer

TABLE 6.2
Predicted response to heat stress (see Table 6.1)

Prediction and interpretation

ISO 7243	ISO 7933				25-node model			time to	time to
WBGT	tcr	tsk	sweat	DLE	tcr	tsk	sweat	confusion	death
°C	°C	°C	litres	mins	°C	°C	litres	mins	mins
32.45	>39	>36	11	26	42	41	3	120	>210

model can be refined to use more detailed information about individuals but will still provide approximations. It should also be remembered that however 'fit' and motivated an individual, it is the physics of heat production and heat transfer that will influence body temperatures and the physiological properties of the body that will determine survival.

Sebastian said that from the results of the computer models he had gained an understanding of the process, its use and its limitations. Also why physiological monitoring may well be required. Actually, the case had not been a hypothetical one and that the officer was a close friend and medical officer that had undertaken the journey with some confidence but had been found in a confused state entering the town in hot conditions and captured by the enemy.

THE REPORT

Advice on procedures to avoid heat casualties during field trials for the selection of special forces personnel.

EXECUTIVE SUMMARY

Objective 1: To advise on instruments and methods for monitoring and measuring the heat stress of the environment as well as criteria for determining the risk of heat injury based upon heat stress

Heat stress is the integrated effect of air temperature, radiation (e.g. the sun), humidity and air velocity on a person when combined with metabolic heat production due to level of activity and the resistance to heat transfer of clothing and equipment. It is recommended that a system of determining the risk of heat casualties is based upon an alert system involving meteorological alert systems; the Universal Thermal Climate Index (UTCI) and the Wet Bulb Globe Temperature Index (WBGT. Actions based upon risk should be agreed and decisions made (e.g. postponement or level of monitoring) by an identified responsible person or persons.

Objective 2: To advise on instruments and methods for monitoring and measuring the heat strain of participants subjected to the heat stress and on the criteria for determining the severity of heat strain including levels that may be considered unacceptable and a danger to health

Heat strain should be monitored at strategic points on the trial, using objective measures including internal body temperature; heart rate and a cognitive performance test. Subjective and behavioural measures may be useful as supporting information but may be unreliable and should not be used solely in decision-making. Actions should be based upon previously agreed criteria, not influenced by the participants in the trial who are subjected to heat stress, irrespective of their rank, and decisions made (e.g. withdrawal) by an identified responsible person or persons.

Objective 3: To advise on methods that can be used to predict the likely heat strain of participants from the heat stress to which they are expected to be, and could be, exposed

Computer simulations are most valuable in testing a range of scenarios concerning heat stress and can predict heat strain. The WBGT index; the predicted heat strain (PHS) model (based upon the body heat equation) and the two-node model of human thermoregulation are presented as examples.

This confidential report provides the first stage of the development of a system for ensuring the avoidance of heat casualties. Methods for the following stages are provided. It is useful to note the acronym SHAFTS as an aid memoir. Sensible (appropriate behaviour) is of particular importance and involves identifying adaptive opportunities for reducing heat strain and training to take advantage of them. It also includes not underestimating the dangers of heat and making wise decisions (not based upon overconfidence). Hydrated; Acclimatised; Fit; Thin and Sober (avoidance of alcohol and drugs), all require full consideration. Once established, training of personnel in the system will be of great importance.

ACHIEVING OBJECTIVES

Measurement and Assessment of Heat Stress

Objective 1: ***To advise on instruments and methods for monitoring and measuring the heat stress of the environment as well as criteria for determining the risk of heat injury based upon heat stress***

HEAT STRESS

The thermal environment that affects people is primarily determined by the interaction of the four environmental variables of air temperature, radiation (in shade and sun), humidity and relative air velocity (air passing over the body, taking account of wind direction and 'walking' velocity). The thermal stress is also greatly influenced by the metabolic heat produced by the activity of the person (highly motivated soldiers) and the thermal properties of the clothing they wear. ALL of those factors shall be considered in defining heat stress and in any assessment of thermal strain. Consideration of one or a subset of those factors is inadequate as it is the interaction of all factors that will affect a person and hence the participants in the trial.

Details of how to measure, monitor and estimate those variables (typically summarised as single number parameters) are provided in Pandolf et al. (1988) and Parsons (2014, 2019 and 2021). I would note that a whirling (to reduce radiation influence) hygrometer with wet and dry bulb thermometers (note that mercury in glass is not now recommended in case of spillage) provides an excellent way of checking air temperature and humidity.

The black globe thermometer can also be used to determine mean radiant temperature (see discussion of the WBGT index below) and a pyranometer can be used to measure direct solar radiation with a strip shield to blot out the sun and measure diffuse radiation from the sky. Specification of measuring instruments is provided in ISO 7726 (2001) and for the estimation of the thermal properties of clothing ISO 9920 (2007). An estimate of metabolic heat production for a given activity can be derived from ISO 8996 (2004). The above references provide a starting point to determining a system of measurement, monitoring and estimation. Measurement, testing and experience will provide a method for determining values specific to the context of the trial.

THE WBGT INDEX

A full assessment would consider all six variables; however, the integrated affects of the environment on people can be represented by the WBGT heat stress index that is accepted worldwide and was originally developed in the USA to avoid heat casualties during military training (Yaglou and Minard, 1957). The WBGT index is used in Army, Navy, Air Force, sport, working and many more contexts and environments (Parsons, 2014). ISO 7243 (2017) provides details of its measurement and criteria for when heat stress is likely to cause unacceptable heat strain, taking into account activity level and clothing.

The WBGT index is measured with a 150 mm diameter, matt black (copper) sphere with a temperature sensor at its centre (globe temperature, t_g); a cylindrical temperature sensor covered in a continuously wetted (muslin) wick (natural wet bulb temperature, t_{nwb}) and an air temperature sensor (in flowing air, shielded from radiation), giving air temperature, t_{a}.

The WBGT sensors are placed at a position where the people would experience the heat stress and the index is calculated from $\mathrm{WBGT} = 0.7\ t_{\mathrm{nwb}} + 0.3\ t_g$ when out of the sun and when in the sun $\mathrm{WBGT} = 0.7\ t_{\mathrm{nwb}} + 0.2\ t_g + 0.1\ t_a$. ISO 7243 (2017) provides a detailed specification of the instrument and Parsons (2019) provides a full discussion.

It is recommended that the WBGT index is measured at strategic positions on the route of the trial to provide an immediate assessment of risk and that it is used as a fundamental part of the heat stress assessment process.

ALERT LEVELS

National and international weather forecasts often provide a system of heat alerts that could be used as a first indication of risk on the selection trial. In the United Kingdom, weather severity is indicated in increasing severity using colour codes through green, yellow, amber and red warnings. Decisions on whether to postpone the trial and levels of monitoring and support could be based upon those that shall be determined for the place and time of the trial.

A useful climate index related to weather is the Universal Thermal Climate Index (UTCI (Jendritzky and de Dear, 2012). An approximation to the index is provided on the internet (e.g. UTCI - Calculator (antonellodinunzio.online) at the time of writing) and values should be calculated for the expected conditions and used to complement the weather alert system. Because of the arduous nature of the trial, an amber warning and above, and a UTCI of greater than 32 °C (strong heat stress) would indicate danger and an appropriate response.

The WBGT index value is calculated as a temperature (°C) and as the trial involves very high levels of metabolic heat production and military clothing and equipment, as well as a high weighting in the WBGT equation, for the natural wet bulb (a wet bulb thermometer cooled by the evaporation of water), alert levels may appear surprising low.

ISO 7243 (2017) provides a starting point for determining WBGT alert levels above which heat casualties may be expected. They depend upon values for the metabolic heat production and the thermal properties of clothing. For very high levels of activity, and military-type clothing, for fit, acclimatised military personnel, a WBGT limit value could be estimated as 23 °C. As a comparison; WBGT alert limits for 'fun runs' have been suggested at 21 °C; for soccer players at 32 °C, with drinks breaks above 28 °C WBGT; and tennis players at 28 °C WBGT albeit in lighter clothing than that worn in military trials. The danger of damage to health due to heat among people in sport is greatly exacerbated by the level of motivation and consequent overconfidence and denial of possible effects that individuals consider may occur to other people but not to them. Observation of alert levels is of particular importance.

Objective 2: ***To advise on instruments and methods for monitoring and measuring the heat strain of participants subjected to heat stress and on the criteria for determining the severity of heat strain including levels that may be considered unacceptable and a danger to health***

HEAT STRAIN INDICATORS

ISO 9886 (2004) provides details and methods for measuring and monitoring physiological heat strain. Internal body temperature; skin temperature; sweat rate and heart rate are the primary indicators. Personal monitoring systems, with telemetry, are possible and have been used to monitor, in real time, the physiological condition of individual soldiers on the battlefield, astronauts, firemen, workers in hot environments and more. Modern versions should be investigated; however, fitting equipment to people will be time-consuming, inconvenient and cause irritation and discomfort.

MEAN SKIN TEMPERATURE

Mean skin temperature (average over the body) and local skin temperature (feet, face, hands) are more suited to indicate cold strain; however, the convergence of skin temperature with internal body temperature can be an indicator of heat strain. I have found the measurement of skin temperature to indicate heat strain to be more trouble than its worth.

SWEAT LOSS

Sweat loss is important in the heat but difficult to monitor. It is related to weight loss and is particularly important when considering hydration.

HEART RATE

Heart rate is an important indicator of physiological strain and a value above 180 beats per minute during exercise and lower when at rest will indicate significant strain. (A maximum value of 180 minus age in years is a rule of thumb safe level.) A measurement at rest and calculated as beats per minute above normal resting level will indicate heat strain (although psychological effects may confound the measure). Heart rate variability and fibrillation are easily measured with simple instruments and are indications of unacceptable strain requiring medical attention.

INTERNAL BODY TEMPERATURE

The most important physiological measure for assessing heat strain is internal body temperature. This indicates the heat storage in the body although the absolute value must be interpreted in the context of the level of activity of the person. A simple infra-red tympanic temperature measurement system may be possible but a second corroborating method (e.g. under the tongue sensor with closed mouth) would be advised. The system of monitoring and measuring internal body temperature should be given full consideration to ensure that it is practical, valid and reliable.

One or more test stations strategically placed around the trial course will allow measures of internal body temperature to be taken and criteria for termination of exposure to be applied.

Cognitive performance and behaviour are indicators of heat strain and can also be measured at the test centres. An appropriate cognitive test will test the ability to reason and make decisions. A standard psychometric test can be used and should be selected (from those of known properties) and tested. A logical reasoning test is often used. A series of questions such as if A > B AND B > C THEN A > C : true or false? A test for confusion will be to ask a series of questions the participant should be able to answer such as what day is it? Who is the prime minister or president? And so on. This sort of test is often used to test confusion of patients in hospitals. Inappropriate behaviour should be used to support other indicators. Behaviour out of character presents a warning, for example, unusually active, quiet or loud.

Criteria for the participant's termination from the trial must be entirely based upon previously agreed objective criteria such as scores on a test as well as physiological indicators. An internal body temperature of over 39 °C for an active subject requires careful observation and over 40 °C, withdrawal from the trial for recovery. For a resting participant, an internal body temperature over 38.0 °C should raise alarms.

A cautionary note is that self-monitoring by those subjected to heat stress may be useful however, simple measurement devices are often unreliable and an independent system administered by trained personnel is required for the assessment of heat strain.

PREDICTING HEAT STRAIN

Objective 3: ***To advise on methods that can be used to predict the likely heat strain of participants from the heat stress to which they are expected to be, and could be, exposed***

THE WBGT INDEX

ISO 7243 (2017) provides a method for predicting the WBGT value from the air temperature, radiant temperature, air velocity and humidity. Parsons (2019) provides details and a computer program for calculating WBGT values, involving the calculation of t_{nwb} and t_g. Predictions of the WBGT values should be made for each distinct part of the trial where thermal conditions are similar, and a time-weighted average taken for the whole trial. The preferred method is measurement and monitoring of the WBGT values rather than predicting values but if only weather data are available then prediction is useful.

PREDICTION OF HEAT STRAIN USING HEAT TRANSFER CALCULATIONS

A heat transfer analysis considers that metabolic heat can be transferred out of the body by a net heat transfer involving conduction (K), convection (C), radiation (R) and evaporation (E). A thermal audit in any thermal environment accounts for the heat transfer (in Watts per square metre of the body surface area – Wm^{-2}) into and out of the body by each of these avenues of heat transfer. By summing the values (positive for heat loss and negative for heat gain), a final value taken from metabolic heat gain provides a prediction of heat storage (S) within the body and hence a prediction of body temperature and sweat rate, two indicators of thermal strain. ISO 7933 (2023) provides a Predicted Heat Strain (PHS) method based upon the body heat equation. A computer program is described in ISO 7933 (2023) and Parsons, (2019).

PREDICTION OF HEAT STRAIN USING A MODEL OF HUMAN THERMOREGULATION

A simulation of the response of the human body provides an indication of thermal strain. For heat stress analysis a two-node model first proposed by Gagge et al. (1971) will be sufficient to predict skin (shell) temperature, internal body (core) temperature, sweat loss and more. The body is represented (the passive system) as two concentric cylinders (core and shell) with appropriate thermal and inertial properties (simulating the tissues of the body), starting temperatures and properties of clothing. A controlling system simulates sweating and changes in skin temperature and dynamic heat transfer calculations provide the changes through time including skin temperature, sweat rate, internal body temperature and more. Prediction of the thermal conditions of a person over the trial is therefore possible from a knowledge of predicted weather data and estimates of metabolic heat production and the thermal properties of clothing. A computer program is described in Parsons (2021).

Example of predicting heat strain using the WBGT, body heat equation and simulation methods.

Consider a three-section trial involving starting uphill for two hours in shade, followed by level ground in full sun for four hours, followed by downhill in full sun for two hours.

Note. 1: It should be noted that the power of prediction using computer models is in scenario testing, as many sets of conditions can be investigated interactively, with possible influence on trial design.

Section 1: Assume that the early morning air temperature (ta) is 15 °C, and in the shade radiant (and globe) temperature is the same as air temperature ($tr = t_g = t_a$). Air velocity is walking speed of 0.5 ms^{-1} and relative humidity is 50%. The clothing worn is estimated as a dry thermal insulation of 0.2 m^2 °C W^{-1} and vapour resistance of 0.02 m^2 kPa W-1. Metabolic rate going uphill with equipment is estimated as 400 Wm^{-2}.

For section 2: t_a = 25 °C; in full sun t_r = 40 °C; relative air velocity increases to 1.0 ms^{-1} and relative humidity is 50%. Metabolic rate is 200 Wm^{-2} and clothing properties remain the same.

For section 3: t_a = 28 °C; t_r = 40 °C; relative air velocity is 0.5 ms^{-1} and relative humidity is 40%. Metabolic rate is 250 Wm^{-2} and clothing properties remain the same.

PREDICTING WBGT

Section 1: Early morning uphill in the shade.

The natural wet bulb is predicted as t_{nwb} = 9.8 °C. Globe temperature is predicted as 15 °C.

In the shade, WBGT = $0.7 \times t_{nwb} + 0.3 \times t_g$ = 11.4 °C WBGT.

For an acclimatised person, metabolic heat production M = 400 Wm^{-2} and military uniform the WBGT limit = 16 °C.

Section 2: Level ground in full sun

The natural wet bulb is predicted as t_{nwb} = 19.1 °C. Globe temperature is predicted as 30 °C. t_a = 25 °C.

In the sun, WBGT = $0.7 \times t_{nwb} + 0.2 \times t_g + 0.1 \times t_a$ = 21.8 °C WBGT.

For an acclimatised person, metabolic heat production M = 200 Wm^{-2} and military uniform the WBGT limit = 18 °C.

Section 3: Late afternoon, downhill in full sun.

The natural wet bulb is predicted as t_{nwb} = 21.6 °C. Globe temperature is predicted as 33 °C. t_a = 28 °C.

In the sun, WBGT = $0.7 \times t_{nwb} + 0.2 \times t_g + 0.1 \times t_a$ = 15.1 + 6.6 + 2.8 = 24.5 °C WBGT.

For an acclimatised person, metabolic heat production M = 200 Wm^{-2} and military uniform the WBGT limit =19 °C.

OVERALL PREDICTION

WBGT = 0.25 × WBGT1 +0.5 × WBGT2 + 0.25 × WBGT3 = $WBGT_{eff}$ = 18.9 °C.

Weighted metabolic heat production = 0.25 × 400 + 0.5 × 200 + 0.25 × 250 = 262.5 W m^{-2}

For an acclimatised person, metabolic heat production M = 262.5 Wm^{-2} and in military uniform the $WBGT_{eff}$ limit = 18 °C

INTERPRETATION

Heat stress is sufficient to trigger an alert to warn against possible unacceptable heat strain.

PREDICTION USING ISO 7933 (2023)

ISO 7933 (2023) uses a body heat equation to calculate the evaporation, and hence sweating, required (Ereq and SWreq) to provide heat balance (no rise in internal body temperature). It predicts heat strain (PHS) in terms of heat storage (rise in rectal temperature) and dehydration as well as predicted exposure times before the heat strain becomes unacceptable for a person fit for work.

The following parameters are used in the prediction: t_a = air temperature °C, t_r = mean radiant temperature °C, Pa = Partial vapour pressure kPa, Va = air velocity ms^{-1}, Met = metabolic rate Wm^{-2}, Icl = clothing insulation Clo, Work = external work Wm^{-2}, Posture: 1 = sitting; 2 = standing; 3 = crouching; imst = static im value; moisture permeability index; Ap = fraction of body surface covered by reflective clothing; Fr = Emmisivity of the reflective clothing; defspeed = code: 1 if walksp entered, 0 if not, Walksp = speed of walking ms^{-1}, defdir = code: 1 if walking direction entered, 0 if not, THETA = angle between walking direction and wind direction degrees, accl = code: 100 if acclimatised, 0 if not, DRINK = 1 for can drink freely; 0 if not, Tre = Final rectal temperature, Swtotg = total sweat rate in grams, Dlimtre = time before rectal temperature limit is reached mins, Dlimloss50 = maximum water loss to protect mean subject, Dlimloss95 = maximum water loss to protect 95% of population, time min = counter for exposure time in the program.(see ISO 7933 (2023) for detail. A computer program is described in ISO 7933 (2023) and Parsons (2019).

Section 1: t_a = 15; t_r = 15; Pa = 0.9; Va = 0.5; Met = 400; Work = 0; Icl = 1.0; Posture = 2; imst = 0.38; Ap = 0.54; Fr = 0.97; defspeed = 0; Walksp = 0; defdir = 0; THETA = 0; accl = 100; DRINK = 1; swtotg = 9615; Dlimtre = 19; Dlimloss50 = 289; Dlimloss95 = 200; tre=38.6 at Time = 480 min.

For this part of the trial, high heat strain is expected due to the uphill climb and high metabolic heat production. This would be unacceptable for working conditions and is sufficient for precautionary action to be taken in the trial.

Section 2: t_a= 25; t_r = 30; Pa = 1.7 ; Va = 1.0; Met = 200; Work = 0; Icl = 1.0; Posture = 2; imst = 0.38; Ap = 0.54; Fr = 0.97; defspeed = 0; Walksp = 0; defdir = 0; THETA = 0; accl = 100; DRINK = 1; swtotg = 3597; Dlimtre = 480; Dlimloss50 = 480; Dlimloss95 = 480; tre=37.7 at Time = 480 min.

The reduction in heat strain is due to flat terrain and some air movement. If sufficiently recovered from the uphill climb unacceptable heat strain should not occur.

Section 3: t_a= 28; t_r = 33; Pa = 2.1; Va = 0.5; Met = 250; Work = 0; Icl = 1.0; Posture = 2; imst = 0.38; Ap = 0.54; Fr = 0.97; defspeed = 0; Walksp = 0; defdir = 0; THETA = 0; accl = 100; DRINK = 1; swtotg = 8293; Dlimtre = 480; Dlimloss50 = 332; Dlimloss95 = 227; tre=37.9 at Time = 480 min.

Downhill provides an increase in metabolic heat production but not sufficient to cause unacceptable heat strain.

Weighted values whole trial: t_a = 23.3; t_r = 27; Pa = 1.6; Va = 0.75; Met = 262.5; work = 0; Icl = 1.0; Posture = 2; imst = 0.38; Ap = 0.54; Fr = 0.97; defspeed = 0;

Walksp = 0; defdir = 0; THETA = 0; accl = 100; DRINK = 1; swtotg = 5388; Dlimtre = 54; Dlimloss50 = 480; Dlimloss95 = 339; tre=38.0 at Time = 480.

Assessment over the whole trial suggests heat strain will occur but if sufficient recovery is allowed from the uphill start, then heat casualties would not be expected if rehydration is effectively carried out.

ASSESSMENT USING THE TWO-NODE MODEL OF HUMAN THERMOREGULATION

After walking uphill for 2 hours the core temperature is predicted to be 40.54 °C with completely wet skin due to sweating. If sufficient recovery is allowed, then after section 2 of four hours on flat terrain, the core temperature is predicted to be 37.6 °C with the body 86% wet with sweat. Section 3, downhill for 2 hours, predicts a final core temperature of 39.24 °C with the body completely wet.

The conditions would be considered unacceptable for a working environment and it is possible that heat casualties will occur. If a weighted average of conditions were used for simulation, then the final core temperature is predicted to be 39.03 °C at 8 hours and maximum sweating occurs throughout the trial.

INTERPRETATION OF THE PREDICTIONS OF MODELS

For the conditions investigated, I would conclude that unacceptable heat strain is possible and that the trial can go ahead on condition that monitoring and support systems are implemented.

THE REPORT

Objective 4: ***To provide the advice in a report marked 'commercial in confidence', with an executive summary that states the aims and objectives, what was done to achieve those aims and objectives and conclusions related directly to aims and objectives (see executive summary above)***

SHAFTS

A useful military adage will aid in providing guidance for the trial. 'Don't be shafted by the heat' (Sensible; Hydrated; Acclimatised; Fit; Thin; and Sober)

The acronym 'SHAFTS' is a useful guide for the avoidance and tolerance of heat strain as follows:

Sensible
Knowledge and training in factors that cause heat strain and how to avoid them is required (e.g. avoid the heat stress and don't go out when it is hot unless you have to. Keep out of the sun, moderate activity, open clothing, etc.). For the example above, enthusiasm and motivation will tempt participants to climb the initial slope too quickly and the sensible approach will be to set a steady pace with sufficient rests and drinking.

Designing adaptive opportunities for the participants and providing training in how to use them is of great importance. If clothing can be designed so that adjustment can release heat for example then heat strain can be reduced. Advice on work and rest, pacing activity-seeking shade, avoid midday sun and so on, all provide adaptive opportunities and require full consideration.

Hydrated
Dehydration is a major threat in terms of heat strain and cognitive performance. Training in a drinking regime is required and should take account of amount of sweating as well as other factors. It is very important that hyponatremia (too much water) should be avoided and is a risk factor in highly motivated people.

Acclimatised
People who have been exposed to heat stress, increase the ability and capacity to sweat and learn how to behave in the heat.

Fit
Fit people have an advantage in the heat as they are 'used to' heat stress from high levels of activity and tolerance to physical strain. Being fit and healthy can reduce the effects of deterioration with age.

Thin
Thin people have a higher surface area to mass ratio that aids heat loss, as well as a reduced workload in movement particularly uphill. This reduces thermal strain when compared with a fat person, for the same heat stress.

Sober
Alcohol as well as drugs, in particular those that influence thermo-regulation, should be avoided.

RECOMMENDATIONS

1. This report is stage 1 of a process that can lead to a tried and tested system for ensuring the avoidance of heat casualties during the field trial for the selection of special forces personnel.
2. Probable heat stress during the trial should be determined using meteorological data and alert systems, calculation of the Universal Thermal Climate index (UTCI and estimates of the WBGT index. The information should be used to assess the level of risk and to determine consequent actions.
3. Heat stress should be measured during all trials using strategically placed WBGT instruments that represent the heat stress to which personnel are exposed.
4. Heat strain should be monitored by the support team using internal body temperature, heart rate and a cognitive performance test. This should be independent of any self-monitoring systems used, should have clear objective criteria, determined, a priori, for actions such as withdrawal from the trial, should have clear identification of authority and responsibility to make the judgement and should not be influenced by the opinion of the soldiers in the trial.

5. Drinking water to avoid dehydration (but not too much) is essential in hot conditions and training and implementation are of highest priority. They are not considered in this report.
6. Computer models of human response to heat provide a useful tool for investigating and simulating how the personnel will respond to predicted and possible heat stress scenarios.
7. A systems approach to developing a procedure should be taken with clear actions, testing on location and full training for staff.

FURTHER READING

Parsons, K.C., 2019, Human heat stress, CRC Press, Taylor & Francis Group, Boca Ratan, London, New York, ISBN 978-0-367-00233-6

STANDARDS

ISO 7726, 1998, Ergonomics of the thermal environment – Instruments for measuring physicalquantities, ISO Geneva

ISO 7243, 2017, Ergonomics of the thermal environment – Assessment of heat stress using theWBGT (wet bulb globe temperature) index, ISO Geneva

ISO 7933, 2023, Ergonomics of the thermal environment – Analytical determination and interpretation of heat stress using calculation of the predicted heat strain, ISO Geneva

ISO 8996, 2004, Ergonomics of the thermal environment – Determination of metabolic rate, ISO Geneva

ISO 9886, 2004, Evaluation of thermal strain using physiological measurements, ISO, Geneva.

ISO 9920, 2007a, Estimation of thermal insulation and water vapour resistance of a clothing ensemble, ISO, Geneva.

ISO 9920, 2007b, Ergonomics of the thermal environment – Estimation of thermal insulation and water vapour resistance of a clothing ensemble, ISO Geneva

ISO 12894, 2001, Ergonomics of the thermal environment – Medical supervision of individuals exposed to extreme hot or cold environments, ISO Geneva

ISO 15265, 2004, Ergonomics of the thermal environment – Risk assessment strategy for the prevention of stress or discomfort in thermal working conditions, ISO Geneva.

7 Case Study
Environmental Ergonomics Survey of a Cold Store

ENQUIRY

Tom was the manager of a large cold store that acted as a storage hub for a supermarket chain. He managed a 24/7 workforce that worked in extreme cold. Work ranged from fork-lift truck operation to manual handling as well as maintenance and cleaning work. There had been complaints of cold and discomfort but the main concern was that health and safety procedures were implemented to avoid unacceptable cold strain and that clothing was appropriate for the work in the cold.

He contacted Seo-ah, who was an environmental ergonomist at the Institute for the Human Environment, for advice on whether workers were experiencing unacceptable cold environments. Tom wanted to ensure that the environments were optimum for the health, comfort and performance of the workers and that conditions were in accord with current regulations and working practices.

THE MEETING

Seo-ah and Tom had a video conference meeting to clarify requirements and context. They agreed to meet at the cold store hub early at an agreed date, prior to conducting a cold stress survey over the day.

On the day of the survey, Seo-ah (with calibrated equipment and single-page questionnaires) met with Tom and Lisa, who was the workers' representative. Experience had demonstrated to Seo-ah that equipment had to be calibrated for the conditions in which it would be used. Wires become stiff, displays cease to work and batteries have a short life if not kept warm. A 'warm' insulated box would be used to store equipment when not in use. She must also carefully control her own exposure and have a supporting buddy (Lisa) out of cold conditions who would monitor her progress.

Tom explained that goods (food and drink) arrived from all over the world in refrigerated lorries and were recorded and stored in the large, sealed, refrigerated warehouse. They were then selected and repacked in refrigerated lorries for distribution to individual supermarkets to meet their demand. The target temperature across the warehouse was –23°C and powerful fans ensured a homogeneous distribution of temperature. Lighting was on a high roof and there were no significant heat sources.

Some fork-lift trucks included heated cabs, but the driver was required to leave the cab on occasions to remove obstructions. Maintenance work was required daily and it sometimes involved tasks involving fine hand work. Cleaning was a problem as the

DOI: 10.1201/9781003401964-9

cleaners were refusing to conduct ‘wet work’. A three-shift system, three teams of 20 cold store workers, was in continuous operation. Of the 60 workers, 50 identified as male and 10 as female. At any one time, up to ten workers would be in the cold store.

A large investment had been made in protective clothing that was maintained and managed by a store person who was also responsible for procurement. Guidance on what to buy was needed and some workers had actually complained of being too warm and sweaty.

Seo-ah explained that she would observe and measure the cold environment, including the air temperature and air velocity at the position of the workers. She would use a thermistor system to measure air temperature and a vein anemometer to measure air velocity as well as childrens’ bubbles (in the form of a wand extracted from a soap solution) and a mobile telephone video recording system to record air movement strength and direction. The conditions should be set to represent those experienced by workers at work.

Lisa agreed to give a questionnaire to all 60 workers over the following week, to be completed in the ‘warm’ locker room after the worker had just left the cold store (Figure 7.1).

Lisa would supervise the distribution and collection of the questionnaire. She would not advise on responses but would note that the purpose was to improve working conditions; that it was their own individual (and not group or social media) response that was required; that it was how they felt, not how they thought the environment would be described; that no questions should be missed and that responses

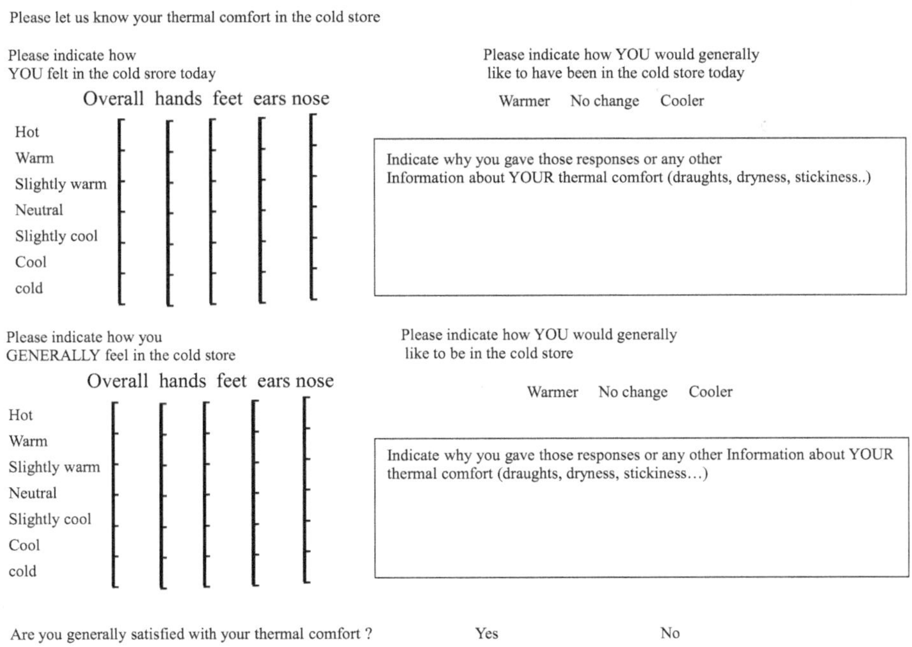
Please let us know your thermal comfort in the cold store

Please indicate how YOU felt in the cold srore today

Overall hands feet ears nose

Hot
Warm
Slightly warm
Neutral
Slightly cool
Cool
cold

Please indicate how YOU would generally like to have been in the cold store today

Warmer No change Cooler

Indicate why you gave those responses or any other Information about YOUR thermal comfort (draughts, dryness, stickiness..)

Please indicate how you GENERALLY feel in the cold store

Overall hands feet ears nose

Hot
Warm
Slightly warm
Neutral
Slightly cool
Cool
cold

Please indicate how YOU would generally like to be in the cold store

Warmer No change Cooler

Indicate why you gave those responses or any other Information about YOUR thermal comfort (draughts, dryness, stickiness…)

Are you generally satisfied with your thermal comfort ? Yes No

FIGURE 7.1 One-page questionnaire for the subjective assessment of thermal comfort in a cold store.

from the previous cold work session and generally at work were required with any additional relevant comments. Identification was needed by initials, time, date and work area, but any individual responses would be kept confidential to Seo-ah, the expert investigator, and not used in any report. A section for clothing (by garment) and type of activity was included as well as personal details (age, gender, etc.).

The questionnaire was shown to Lisa and Tom. Responses to the environment included a rating for how the worker felt in the previous cold work session and generally on the scale (extremely cold; very cold; cold; cool; slightly cool; neutral; slightly warm; warm; hot, very hot; extremely hot) and on the same scale how they generally felt when they were in the cold store. Ratings were provided overall and for hands, face, ears and feet.

Open questions included suitability of clothing, main causes of discomfort, problems with near accidents and any illness or injuries.

Tom was requested to provide any selection and screening criteria for people who work in the cold, particularly related to health. Seo-ah noted that she would also measure and assess conditions in the locker room and rest areas.

AGREED AIM AND OBJECTIVES

Aim

To conduct an environmental ergonomics cold stress survey of the cold store and advise on actions that would ensure that workers did not experience unacceptable cold strain

Objectives

Objective 1: To measure the thermal environment in the cold store and provide advice to ensure that work conditions are within accepted limits to ensure the avoidance of unacceptable strain

Objective 2: To determine the level of thermal strain currently experienced by the cold store workers and to recommend actions to avoid unacceptable cold strain

Objective 3: To provide advice on the selection of clothing appropriate for work in cold conditions

ACHIEVING OBJECTIVES

Measurement and Assessment of Cold Stress

Objective 1: To measure the thermal environment in the cold store and provide advice to ensure that work conditions are within accepted limits to ensure the avoidance of unacceptable strain

Seo-ah used a blueprint map of the cold store and selected 10 positions across the large store that would represent a distribution of workplaces. At each place, she measured air temperature and air velocity and noted the activity and clothing of the workers. She also observed the behaviour of workers to identify adaptive opportunities and methods of reducing cold strain. After one hour, she left the cold store area and measured the air temperature, radiant temperature (with a black globe thermometer), humidity (with a whirling hygrometer) and air movement (using a hot wire anemometer and bubbles) in the locker room and the rest area. She returned to the cold store after one hour of rest and repeated the measurement process.

The dry bulb temperature of the whirling hygrometer is air temperature, the wet and dry bulb temperatures together provided relative humidity and partial vapour pressure using a psychrometric chart and the temperature of the black globe was corrected for air temperature and air velocity to provide mean radiant temperature. Activity and clothing were noted for the two areas. ISO 11079 (2007) was used to determine international standard conditions for the cold store and compare them with existing conditions. ISO 15743 (2008) was used to estimate risk and for advice on working practices. ISO 8996 (2004) was used to estimate metabolic heat production from activity and ISO 9920 (2007) to estimate clothing insulation of garments and ensembles. ISO 12894 (2001) was used to provide information on the screening of workers for work in the cold. ISO 7730 (2005) was used to assess the locker room and rest areas. (Note that the wet bulb temperature was not relevant inside the cold store because of freezing, with low humidity assumed – possibly leading to dryness of skin, lips and eyes especially in wind.)

Measurement and Assessment of Cold Strain

Objective 2: To determine the level of thermal strain currently experienced by the cold store workers and to recommend actions to avoid unacceptable cold strain

Thermal strain of workers was determined using subjective responses from the questionnaire. Scales were developed from ISO 28802 (2012). Physiological indicators of internal body temperature, local skin temperatures and heart rate may be required in a follow-up study if necessary. Actions to avoid unacceptable cold strain were derived from the ISO standards cited above for objective 1.

Selection of Clothing

Objective 3: To provide advice on the selection of clothing appropriate for work in cold conditions

Clothing was considered in terms of the overall clothing insulation required for the work; protection of local exposed areas of the body (with large surface area compared to their mass or volume, e.g. fingers, hands, toes, face, ears, nose, etc.) and the requirements of the work as well as any adaptive opportunities (hands in pockets; clothing adjustment). ISO 11079 (2007) provides a method for calculating clothing insulation required and ISO 9920 (2007) on how to achieve it. ISO 15743 (2008) provides advice on working practices for work in the cold.

A follow-up meeting was held at the end of the day where all was confirmed as according to plan and that Lisa would send the completed questionnaires, via Tom, after one week. It was agreed that the report, marked 'commercial in confidence' would be sent to Tom at an agreed date and within one month.

THE REPORT: AN ENVIRONMENTAL ERGONOMICS SURVEY OF A COLD STORE

Executive Summary

An Environmental Ergonomics survey of a cold store (West Terrace site) was conducted on 19th September 2024. The aim was to determine existing cold stress and strain and advise on actions that would ensure that workers did not experience unacceptable cold strain. Cold stress was measured across the cold store. Air temperature was found to be between −22 °C and −24 °C (with a target of −23 °C) and air velocity varied with the control system between 0.2 ms^{-1} and 1.0 ms^{-1}. Mean radiant temperature was equal to air temperature and humidity in cold air can be considered to be very low potentially leading to dry lips and skin. Clothing insulation ranged from an estimated 1.5 clo to 2.3 Clo and activity level ranged from standing at rest (65 Wm^{-2}) to heavy manual work (150 Wm^{-2}). International standard ISO 11079 (2007) indicates that for an air temperature of −24 °C and air velocity of 1.0 ms^{-1}, for people standing at rest, clothing insulation required is not achievable and conditions are unacceptable for continuous exposure (8-hour day). Cold discomfort overall, and in local areas of the body, were found as well as some dryness.

For conditions required in the cold store with manual work, acceptable exposure times in existing clothing range from 50 to 108 minutes if fans can be switched off during work. It is recommended that work should be redesigned to involve a 1-hour exposure rotation system and appropriate clothing with minimum garments selected for an ensemble that is added to by individual worker selection, particularly of clothing for extremities.

A clear plan of work should be established that involves continuous work and avoids waiting and resting in the cold store. The feasibility of control system operation, (especially the use of fans) to avoid significant air movement across workers at work times, should be investigated, coordinated with a work-rest system for workers and advised by methods provided in ISO 11079 (2007). Locker room and rest areas were in line with standards for thermal comfort (ISO 7730, 2005).

RESULTS OF THE SURVEY BY OBJECTIVES

Objective 1: To measure the thermal environment in the cold store and provide advice to ensure that work conditions are within accepted limits to ensure the avoidance of unacceptable strain

Across the cold store, the refrigerating systems maintained air temperature at an average (mean) of −23.2 °C with a range of −22 °C to −24 °C. Air velocity ranged from 0.2 ms^{-1} to 1.0 ms^{-1} in intermittent bursts (to maintain air temperature at target levels). ISO 11079 (2007) uses a method of calculating the heat exchange between a clothed worker and the cold environment to estimate the clothing insulation required (IREQ) to ensure acceptable working conditions. The most extreme cold conditions in the cold store are an air temperature of −24 °C air (= mean radiant) temperature, 1.0 ms^{-1} air velocity, dry air and a clo value of 1.5 clo while workers were waiting to work (standing at rest with an estimated metabolic heat production of 65 Wm^{-2}).

For those conditions, the minimum required clothing insulation is estimated around 7.0 Clo which is not achievable in practice for work clothing. At 1.5 Clo, the duration limited exposure (DLE) is limited to 12 minutes for comfort (DLEneutral) and 18 minutes, where workers would feel uncomfortably cold but it is not dangerous (DLEmin). The method used is limited in accuracy for such short-time predictions and recovery time to comfort (RTneutral) is not specified as the conditions are clearly unacceptable. If the fans were not blowing across the workplace, at the specified −23 °C and a low air velocity of 0.2 ms^{-1} and work at 150 Wm^{-2}; the IREQmin would be 2.3 clo (2.7 clo if air penetrates the clothing e.g. through openings), the DLEmin is 96 minutes and the DLEneutral is 54 minutes.

It would seem reasonable to plan work when fans are off and for periods of 1 hour with at least 1 hour outside of the cold area for recovery maybe working in thermally comfortable conditions. Work redesign may require a rotating system and my experience has been that once people had worked for a period in the cold they did not want to return to it even after rest. Putting on and taking off clothing is often a cause of dissatisfaction. Sweating should be avoided so work should be paced and avoid peaks in activity. Toilet breaks should be anticipated in cold conditions.

The wind-chill index and wind-chill temperature are both used to determine human response to cold conditions and provide guidance, particularly for local cooling of the body (ISO 11079, 2007; ACGIH, 2024). For an air temperature of −23 °C and an air velocity of 1.0 ms^{-1}, the wind chill index is around 1000 with a chilling temperature of −12 °C corresponding to a feeling of very cold. The more recent wind chill index (wind chill temperature – ISO 11079, 2007) provides a value of twc = −22 °C corresponding to a feeling of uncomfortably cold and moderate risk with Environment Canda suggesting dress in layers of warm clothing with a wind resistant outer layer. Wear hat, mittens, scarf, insulated waterproof footwear and stay dry and active (Parsons, 2021).

Limiting values are often taken as 1400 for the wind chill index where exposed flesh may begin to freeze and a wind chill temperature around −24 °C for 1 ms^{-1} air velocity and −30 °C for calm conditions. The wind chill indices are however mainly developed for use in outside conditions but are useful in the current context as supporting information.

Objective 2: To determine the level of thermal strain currently experienced by the cold store workers and to recommend actions to avoid unacceptable cold strain

The subjective questionnaires and comments indicated that workers were generally satisfied with the working conditions even though they were uncomfortably cold on occasions. A consistent problem was the lack of coordination and planning such that some workers could stand around for 10 minutes or more in the cold store with no work. For those cases, they attempted to keep out of the wind but soon had very cold hands and feet and were glad when work started 'so that they could warm up'.

There was no concern over levels of accidents injuries or absenteeism. There were no obvious differences between male and female responses. It is recommended that the cold store environment be set as for objective 1 and that the work be redesigned to address ratings and comments made in the questionnaires.

For each worker, clothing should be provided to a minimum of 2.0 clo and then workers should select further clothing, particularly for extremities (hats, gloves, etc.) to meet their individual requirements. Workers should be involved in the design of the work, clothing and also asked for ideas to improve working conditions.

Objective 3: To provide advice on the selection of clothing appropriate for work in the cold conditions

An analysis of the cold conditions using ISO 11079 (2007) indicates that clothing with an insulation of 2.3 clo and normal vapour permeation properties would provide a minimum acceptable clothing insulation and that additional garments will protect workers from being uncomfortably cold towards thermal comfort. ISO 9920 (2007) provides a database of the thermal properties of clothing garments and ensembles that will allow selection of appropriate clothing.

A common standard clothing ensemble of 1.0 clo insulation should be worn by workers outside of the cold store. An additional cold store jacket and trousers should be worn for work in the cold store, to increase the ensemble clothing insulation to above 2.0 clo. Boots and thermal socks should be worn and a selection of additional garments should be provided to include protection for the hands, head, face and any exposed skin. The additional items should be selected by the workers to suit their requirements and experiences.

SUMMARY OF MAIN RECOMMENDATIONS

1. Clothing ensembles should consist of a basic ensemble of around 1.0 clo insulation added to by insulated boots, a hat and gloves and a cold store jacket and trousers taking the clothing insulation of the ensemble to 2.3 clo.
2. Additional garments for the avoidance of local discomfort should be made available so that individual workers can select according to their requirements.
3. The required temperature of −23 °C should be maintained by the minimum use of fans that influence workplaces, to avoid airflow over workers.
4. Work should be redesigned with careful planning to avoid waiting periods and typically for 1-hour periods with a maximum of 90 minutes exposure. If work must take place in moving air then maximum exposure should be 30 minutes.

ADDITIONAL RECOMMENDATIONS

5. Workers should adjust clothing and pace activity level to avoid sweating.
6. Maintenance work involving fine dexterity will require special attention to avoid cold hands. Heated gloves and heated pockets may be required.
7. Cleaning using wet products should be avoided. It is unacceptable at –23 °C air temperature, particularly with wind chill.
8. When adjustments have been made and a new system implemented, monitoring and evaluation will be required in an iterative process to optimise the procedures.

FURTHER READING

Parsons K.C., 2021, *Human cold stress*, CRC Press, Taylor & Francis Group, Boca Ratan, London, New York, ISBN 978-0-367-55199-5

STANDARDS

ISO 7726, 1998, Thermal environments – Instruments and methods for measuring physical quantities, ISO, Geneva

ISO 8996, 2004, *Ergonomics of the thermal environment – determination of metabolic rate*, ISO, Geneva

ISO 9920, 2007; *Estimation of thermal insulation and water vapour resistance of a clothing ensemble*, ISO, Geneva.

ISO 11079, 2007, *Ergonomics of the thermal environment – Determination and interpretation of cold stress when using required clothing insulation (IREQ) and local cooling effects*, ISO, Geneva.

ISO 12894, 2001, *Ergonomics of the thermal environment – Medical supervision of individuals exposed to extreme hot or cold environments*, ISO, Geneva.

ISO 15265, 2004, *Ergonomics of the thermal environment – Risk assessment strategy for the prevention of stress or discomfort in thermal working conditions*, ISO, Geneva.

ISO 15743, 2008, *Ergonomics of the thermal environment – Cold work places – Risk assessment and management*, ISO, Geneva.

Part III

Environmental Ergonomics and the Assessment of Noise and the Acoustic Environment

Part 3 presents the principles of human response to sound and provides examples of how to conduct an environmental ergonomics assessment of an acoustic environment. It is in four chapters. Chapter 8 describes the nature of sound; the measurement of sound and the human response to sound. A case study investigates noise in a papermill (Chapter 9) and Chapter 10 considers the effects of noise on annoyance, communication and task performance. An environmental ergonomics assessment of the acoustic environment in a call centre is presented in Chapter 11.

The big bang at the start of the universe was silent as sound needs matter to exist. The same is true on the moon where there is no atmosphere. On earth, people have evolved to detect sounds and Part 3 describes how systematic research and practical application have led to methods for the assessment of acoustic environments and how sound will affect people who occupy them.

The ear is the specialist organ that detects sounds and two of them spaced apart on the head allow the ability to detect the position of sound sources. We should credit scientists, physicians and physiologists as well as mathematicians for an understanding of sound perception. Thomas Willis as early as 1672 speculated that the ear detected tones using fibres in the cochlea. Fourier showed how complex waves can be separated into component sine waves and Ohm suggested that this was how the ear analysed sounds. This was developed by Helmholtz into a theory of hearing. Much development has taken place since then especially involving the nervous

DOI: 10.1201/9781003401964-10

system, brain and signal interpretation. A schematic of the system of hearing is shown in Figure P3.

The physical level of sound (pressure) is not directly related to the human response. Bel laboratories proposed the decibel as a unit related to telephone communication in 1922. Sound is related to the psychophysical Weber-Fechner law. This stated that the just noticeable difference from one level of sound to the next is related to the original level of sound. Therefore, it takes more sound to achieve a just noticeable difference at higher levels.

This provides a logarithmic relationship between human perception in terms of loudness for example and the physical level of the sound. So the decibel, as a relative logarithmic function provides the basis for human response. A rule of thumb is that for every increase of 10 dB, loudness is doubled. This is developed into units and indices used in environmental ergonomics assessment of the acoustic environment.

Chapter 8 presents fundamental principles for the detection and measurement of sound and the measurement and prediction of the human response to sound. Sound is detected as a pressure wave in the ear that is detected, transmitted to and interpreted by the brain. Noise is unwanted sound. The decibel is defined as a measurement of sound, measured using a sound level meter.

Equal loudness curves are presented that show the relative loudness of different frequencies of sound. The inverse of those curves provides idealised weighting functions that are used in units to predict human response, such as annoyance. The dB(A) level is an example and when represented over an 8-hour day provides the dB(A) Leq value. Limit values are internationally at levels around 85 dB(A) Leq for an 8-hour day and provide action limits for wearing hearing protection and the avoidance of permanent hearing loss.

In Chapter 9, an internal investigation into noise in Hall 5 of a papermill was requested by Mohamed, the chief medical officer. Environmental Ergonomist, Henry,

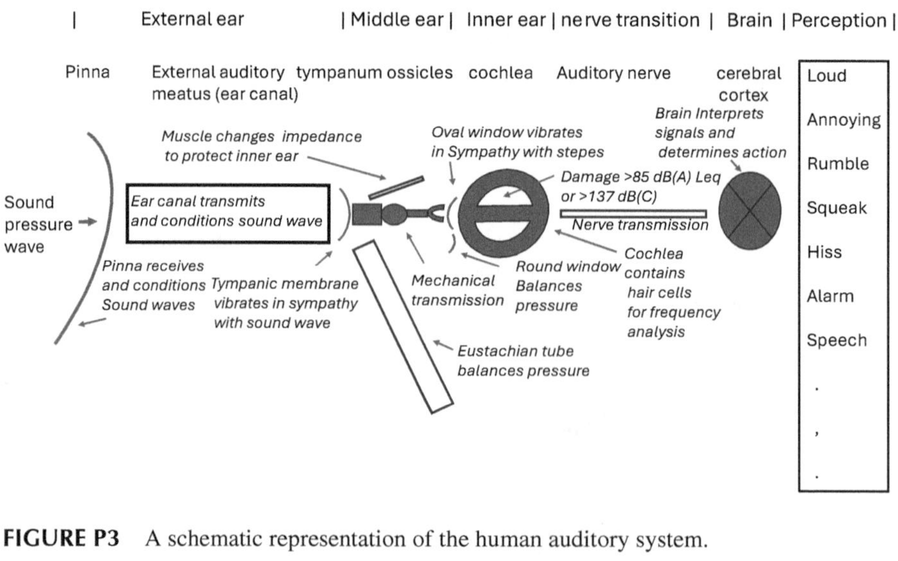

FIGURE P3 A schematic representation of the human auditory system.

and his team conducted a noise survey to produce a noise map in Hall 5; estimate daily exposure times for workers; determine the effects of noise on the workers; and compare levels of noise in Hall 5 with accepted national and international limits.

For the measurements of noise, subjective responses of the workers and observation checklists concluded that hearing protection was required throughout the Hall. It was concluded that further improvements could be made. Health screening should be improved; possible reduction of noise at source and reduction of exposure of workers should be considered; warning signs should be displayed and consideration should be given to employing noise reduction specialists.

Noise discomfort is often expressed as annoyance and a multitude of indices are available for its assessment and some are described in Chapter 10. Measurement of dB(A) level is commonly used in indices for different contexts. It is concluded that maps of dB(A) levels will provide a useful first step in any environmental ergonomics survey. Predicting the effects of noise on performance is considered in the context of the RDC (Regulations; Distraction; Capacity) model. Speech interference, communication, warning, alarm and information systems are also considered as well as background noise and signal-to-noise ratios.

Chapter 11 presents an environmental ergonomics survey of noise in a call centre in a large open-plan office. Paul, an Environmental Ergonomist at Fukoyo University, conducted a noise survey at the request of Haruto, manager of the centre. A noise map was produced of the office using dB(A) measurements, and subjective and observational methods measured worker responses. Measurements of noise were compared with guidelines for offices and for call centres.

Four shifts were assessed and average results showed an average of 65, 70, 67 and 63 dB(A) values for morning, afternoon, evening and night shifts, respectively. Recommendations included removal of a printer from the office; more effective screening of workers, especially after work; and training in voice level control for some telephone operators. A target level should be set as a maximum of 60 dB(A) for the call centre.

8 Human Response to the Acoustic Environment

HUMAN RESPONSE TO THE ACOUSTIC ENVIRONMENT

Mechanical energy from moving objects causes pressure changes in the fluid in which they operate. The pressure changes travel as waves (longitudinal – compression and rarefaction along the direction of travel), away from the object. People detect and interpret those pressure changes through a specialist acoustic system involving detection by the ear and mechanical and neurological transmission to the brain where it is perceived and interpreted as sound.

The pressure changes can affect the health, comfort and performance of people and an environmental ergonomics survey will provide an indication of the nature and degree of those effects on occupants of an environment.

As well as the health, comfort and performance of the occupants of an environment, it should be remembered that, although important, this can be a simplistic consideration. The acoustic environment (as well as other environmental components and their integration) can also influence the aesthetics of an environment involving emotions and moods as well as pleasure, stimulation and the consequence of interpreting acoustic stimuli, including speech, signals, music, concern, alarm and more.

SOUND

Most of the mechanical energy interpreted as sound is received at the ear and is caused by changes in air pressure with time. The pressure at any point in time is called the sound pressure level (SPL) and the change in SPL with time provides sound waves. So sound can be described by its pressure level (SPL; 1 Pa = 1 Nm^{-2}), wavelength (λ; distance over one complete cycle in time: metres) and frequency (f; how many wavelengths are completed in one second: 1 Hz = 1 cycle per second).

Consequently, the velocity of the sound is $v = f\,\lambda$, ms^{-1}. For completeness, we should add that in air at a temperature of 20 °C, the speed of sound is 344 ms^{-1} and that the energy flow (power or intensity) is related to the square of the pressure.

As the level of air pressure changes with time, a root mean square (rms) value is taken to represent the overall sound level over a period of interest. If actual pressure levels were used as units to assess sound, a very wide range of values would be required. For reasons of convenience, that people roughly perceive sound level as a logarithmic relationship between pressure and perceived intensity; and to reduce the range of units to manageable proportions, the decibel unit is used to represent sound level.

The absolute threshold of sound pressure (P_o) at a frequency of 1000 Hz is standardised as an air pressure of 2×10^{-5} N m^{-2}. The intensity is related to the square of the pressure so the decibel is calculated as $10 \log_{10} (P^2/Po^2) = 20 \log_{10} (P/P_o)$ where P is the sound pressure being measured.

DOI: 10.1201/9781003401964-11

The environmental ergonomist should be clear about what this means. If sound is at the hearing threshold, then P/P_o is 1 and $1 = 10^0$ so 20×0 is 0 and we have 0 dB for the hearing threshold. If P/P_o is 100, then we have $100 = 10^2$ and $20 \times 2 = 40$ dB and so on. Haslegrave (2015) gives examples as 0 dB for the hearing threshold; 65 dB, for conversation; 80 dB, for town traffic; 100 dB, for workshop noise; 125 dB, for a jet at takeoff and 140 dB as the threshold of pain.

If we double the sound intensity (then rounding $2^{1/2}$ to 1.4), $dB = 10 \log_{10} (2P^2/P_o^2) = 20\log_{10} 1.4P/P_o = 20\log_{10} P/P_o + 20\log_{10} 1.4/P_o$ = the original dB value plus $20\log_{10} 1.4/P_o$. As $10^{0.15} = 1.4$; and $20 \times 0.15 = 3$ dB, doubling the sound intensity increases the dB value by 3 dB. Haslegrave (2015) gives +3 dB = 1.4 × SPL = 2 × sound power level; +6 dB = 2 × SPL = 4 × sound power level and +20 dB = 10 × SPL = 100 × sound power level.

It follows from the above discussion that two sounds of equal sound power level would combine to add 3 dB to the overall sound power (e.g. 75 dB + 75 dB = 78 dB). In general, for two sound sources, the source with the higher power is added to by the sound with the lower power by 3 dB if it is 0 dB lower; 2.5 dB if it is at 1 dB lower; 1 dB if it is 6 dB lower and 0 dB if it is 10 dB or more, lower.

It is important to remember that the dB value discussed so far is a representation of the physical energy and is not adjusted for the characteristics of the human auditory system. As this is a linear relation without adjustment it is sometimes termed dB(lin). This value represents the physical sound that can be measured in an environmental survey. To predict the effects of the acoustic environment on people, an adjustment must be made for the human sensitivity to the level and in particular to the frequency of the sound. The physical measures are often referred to as intensity and frequency and for the corresponding perception by people, loudness and pitch.

HUMAN RESPONSE TO SOUND

Noise is unwanted sound and is therefore personal to an individual. One person's noise can be another person's sound which is 'not unwanted'. Background noise can be considered to be the unwanted sound, in the background, not relevant to the activity being conducted. Loudness is a subjective term and the relationship between the level and frequency of sound and loudness has been determined using human subjects to establish equivalent loudness when exposed to a range of sounds of different levels and frequencies.

The results of the subjective studies in a laboratory allow equivalent loudness curves to be constructed where the equivalent level in dB of sounds over the frequency range of 20 Hz to 15 kHz are plotted for standard levels in dB at a frequency of 1000 Hz. For example, a 50 Hz tone (single frequency sine wave) requires 85 dB to give equivalent loudness as a 1000 Hz tone at 50 dB. This implies that people are 1.7× (at that level) more sensitive in terms of loudness at a frequency of 1000 Hz than they are at 50 Hz (Figure 8.1).

The relative sensitivity to sound frequency changes with sound level and for any sound frequency, the equivalent loudness level at a frequency of 1000 Hz provides a single scale with sound level and is called the Phon unit in dB. The Phon scale,

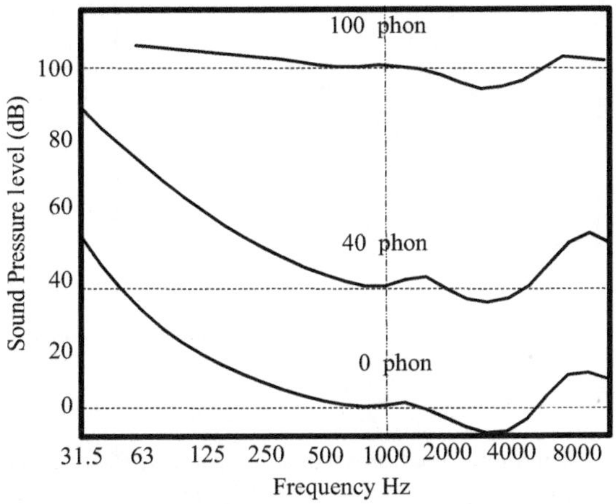

FIGURE 8.1 Equal loudness curves for pure tones. The 0-phon curve (0 dB at 1000 Hz) is close to the threshold for hearing. The 40-phon curve (40 dB at 1000 Hz) was used to derive the A-weighting function used in the derivation of dB(A) and the 100-phon curve (100 dB at 1000 Hz) is used for weightings related to high intensities (dB(C)). The inverse of the curves can be compared with Figure 8.2.

however, is not linear so for example 43 (dB) Phons is not twice as loud as 40 Phons. For this reason, the Sone scale was developed where Phon = 40 + 10 $\log_2$ Sone or Sone = $2^{((\text{Phon}-40)/10)}$. So a loudness of 40 Phon is $2^0 = 1$ Sone; 50 Phon = $2^1 = 2$ Sone so twice as loud; but 60 Phon is $2^2 = 4$ Sone so 4× as loud.

Equivalence of subjective pitch is provided by the Mel unit defined as the equivalence in perceived pitch as a pure tone frequency of 1000 Hz at a Sound Pressure Level (SPL of 60 dB. Any two tones separated by a given number of mels appear equally apart in pitch regardless of their frequency (Haslegrave, 2015). I have not used the mel in environmental ergonomics application but the equivalent loudness curves are widely used in the form of frequency weighting curves and are an integral part of the assessment of the acoustic environment.

Sound can be represented as a pure tone with a single frequency with a sinusoidal representation in the 'time domain' (level vs. time graph) and a single value on a spectrum in the frequency domain (e.g. level vs. frequency graph).

It can also be represented as a complex wave with irregular pattern in the time domain. Fourier analysis can represent any complex wave into its frequency components and in the frequency domain into a combination of sinusoidal waves of different frequencies with appropriate levels in a spectrum. As Fourier analysis also preserves the phase relationships (real and imaginary parts in mathematics) between the sine waves, an inverse Fourier 'transform' will reconstruct the complex wave back into the time domain. Measurement and analysis systems are widely available so the environmental ergonomist has to understand the values but does not have to calculate them.

Frequency weighting curves that describe the average relative sensitivity to sound frequency of the human auditory system are integrated into measuring and analysis systems to provide weighted decibel values. Three internationally standardised

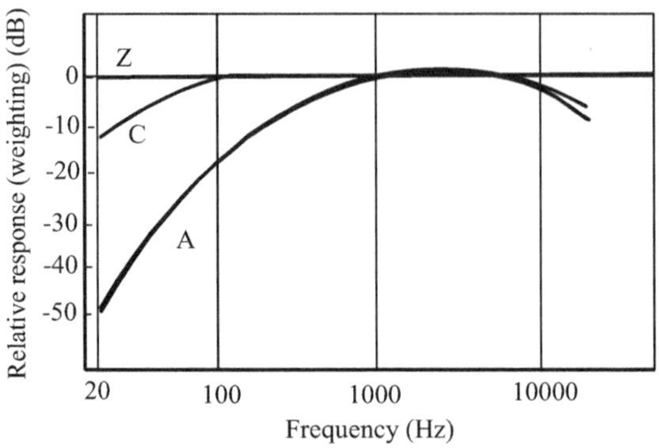

FIGURE 8.2 The A weighting function used to determine dB(A); the C weighting function used to determine dB(C); and the Z weighting function used to determine dB(lin).

weighting curves are A, C and Z, for low, moderate and high sound levels respectively (see Figures 8.2). The A-weighting curve, for example, is based upon the inverse of the 40 dB Phon curve, interpreted as sensitivity instead of loudness.

The principle is that a weighting of 1 (that is a weighting of 0 dB) provides the multiplication factor for the level at a frequency of 1000 Hz and levels at other frequencies are given a value calculated from the relative sensitivity of that frequency of sound to that at 1000 Hz. For example, the 50 Hz example given above would have a weighting of 1/1.7 (approx. 0.6 or in the 'land of the decibel', –35 dB).

A weighting curve can then be multiplied by respective levels at frequencies in a measured spectrum to provide a weighted spectrum that when integrated over all frequencies will provide a weighted value.

The unweighted level (weighted with a value of 1 across all frequencies is the Z-weighting) provides dB(lin) and weighted values are given as dB(A) and dB(C). (dB(B) was based upon the 70 Phon curve but is not now used; dB(D) gives more weight to higher frequencies and is used for the assessment of aircraft noise). So weighted values are equal to dB(lin) for a pure 1000 Hz tone but mostly lower where other frequencies are present. dB(A), for example, is often used to assess offices in environmental ergonomics.

If significant low-frequency sound is present, then dB(A) will be less than dB(lin) as it is assumed in the A-weighting that low-frequency sound is less intrusive than sounds at a higher frequency. It is also assumed that equal-weighted values for any complex wave give a valid and equivalent comparable indication of the affects of the sound or noise on a person. For example, 65 dB(A) can represent a wide range of sounds of different characteristics but they all give equivalent levels of annoyance.

MEASUREMENT OF THE ACOUSTIC ENVIRONMENT

Environmental ergonomics surveys are often concerned with measuring noise (unwanted sound) that will vary across a space in time and level (intensity) and characteristics (e.g. frequency) often with multiple sources and influenced by architecture and design of the space as well as its contents.

Statistical procedures are used to describe the parameters of the sound and represent its time-varying characteristics. If we consider a sound over time, then L_n is the level that is exceeded n% of the time. The *median noise level* is L_{50}, the level of noise that is exceeded 50% of the time. The *background noise level* is L_{90}, the level of noise that is exceeded 90% of the time. The *peak noise* level is often given as L_1 or sometimes L_{10}, exceeded 1% or 10% of the time, respectively. Other parameters can be useful, for example, a range can be described as L_{90} to L_{10} and so on (Figure 8.3).

The equivalent energy over a period of time has been shown to be related to effects on people and in particular to hearing loss (ISO 1999, 2013). It integrates variations in noise level over time so, for example, for a machine operating intermittently over a day, a single equivalent energy value can be calculated as if it had been a continuous single-level noise over the day. This is termed the *equivalent level of sustained noise* (Leq). Haselgrave (2015) describes it as the average level of sound energy over a given period of time, which integrates all of the fluctuating noises to represent them as an average steady level. It, therefore, takes account of high peaks of short duration as well as low inactive levels over relatively long durations.

All of the above parameters can be determined from sound time histories and require a specified period of time. They are also dB(A) weighted values (although technically they can be calculated without weighting). So, for example, an Leq can be described as 8-hour Leq, dB(A). If we use 85 dB(A) Leq as a limit for industrial work, then if exposure is for 4 hours, 88 dB(A) Leq will be the limit, for 2 hours exposure, 91 dB(A), Leq and so on (remember 3 dB doubles the sound intensity).

The control of noise at work regulations (HSE, 2005a) is based upon European regulations and sets 80 and 85 dB(A), Leq as action limits (lower and upper) for a daily exposure at work and 135 and 137 dB(C) for peak noise. Exposure limits shall not be exceeded and are 87 dB(A) daily or weekly personal noise exposure with 140 dB(C) peak sound pressure, that is, taking account of the characteristics of any hearing protection (ear muffs, ear plugs and noise cancellation devices). The protection should be worn correctly and will have a frequency dependency often with low attenuation to low-frequency sound and higher attenuation at higher and potentially damaging sound frequencies.

The *day-night equivalent level (L_{dn})* can be used to assess community noise and is a 24-hour dB(A) Leq with the 'night-time' measures between 2200 and 0600

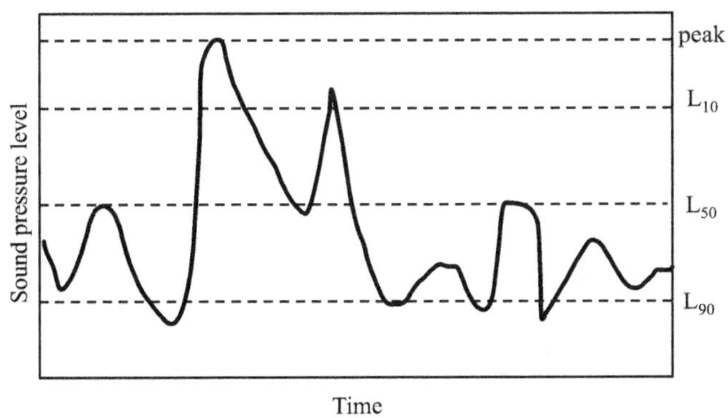

FIGURE 8.3 Sound time history showing L_{10}, L_{50}, L_{90} and peak levels.

weighted with the addition of 10 dB. The *sound exposure level (SEL or L_{AE})* integrates energy over the duration of an event and is expressed as the equivalent level over 1 second. It is useful for comparing single events such as a passing car or train or for comparing unrelated events.

Frequency analysis of noise, plots level of sound with frequency and is useful to identify the components of the noise in terms of sources. If an unexpected frequency component is present, it can be used to identify a problem in machine operation for example or it can be used to identify the relative contribution of a sound component to an outcome such as hearing loss in workers. Original frequency analysis used octave (doubling of frequency) or third octave (logarithmically divide octaves into three) frequency bands. More recent analysis uses finer frequency bands in digital analysis. The most common analysis is to plot a frequency or power spectrum.

NOISE MEASUREMENT

Noise is measured with a sound level meter (EN 61672-1, 2003), which includes a microphone that detects sound pressure changes and converts them into electrical signals that can be recorded and analysed to provide representations in both time and frequency domains and determine the parameters described above and more.

Correct operation of the sound level meter is essential to provide the required measurement and in environmental ergonomics that is often the sound received at the ear by people across a space and over time. It must not be contaminated by the person making the measurement, so a microphone stand is often used.

A noise survey will require a design that is specific to meet the requirements and context of the survey and may involve supporting measurements of background noise without occupants; of specific machinery in anechoic or reverberant chambers in a laboratory; and more. Difference techniques (machines on, then off, etc.) may be useful in attempting to understand the acoustic environment to which people are exposed.

Dosemeters are small microphones attached to people that integrate sound level over time to provide a total exposure or dose, usually used to calculate dB(A) Leq. Probe microphones and artificial ears are used to investigate noise in the ear often to investigate the effectiveness of ear protectors. Microphones must be protected from low-frequency noise caused by outdoor wind using open cell foam. Sound-level meters are calibrated in the field, using a pistonphone that fits over the microphone and produces a single tone (e.g. 94 dB at 1000 Hz) signal that can be compared with the value given on the sound level meter display.

Numerous national, regional and international standards are available reflecting the importance of the acoustic environment worldwide as well as its complexity. ISO 3740 (2000) provides general guidance and ISO 28802 (2012) provides guidance on its role in the environmental ergonomics survey. HSE (2005b) provides practical guidance including how to conduct surveys in industry and relate the results to worker health.

The USA provides noise limits at the workplace as a permissible exposure limit (PEL) of 90 dB(A) Leq for an 8-hour day, in standard OSHA 1910.95. It is important to note that a 5 dB ‘exchange rate’ is used where the allowable exposure time is halved for every increase of 5 dB in noise level, which contrasts with an exchange rate of 3 dB often used internationally. The ACGIH (2024) provides threshold limit values for continuous or intermittent noise over a day of 85 dB(A) for 8 hours, 88 dB(A) for 4 hours exposure and so on. They specify using a standard sound-level meter with dB(A) filters and a slow meter response (providing a longer integration time). For outside exposure, a value of 70 dB (A) is often taken as a limit and at night 34 dB(A) or 10 dB(A) above background noise.

Very loud noises often of short duration can cause pain and even damage to the ear, and high levels of low-frequency sound (infrasound) can vibrate the whole body with mild discomfort, or sometimes alarm and even death as pressure waves damage internal organs (e.g. caused by explosions). For completeness, we should mention ultrasound that can penetrate the skin and is used to show images of bodily structures (e.g. developing babies or damage to tissues). In between the extremes, noise can cause long-term damage to hearing, and at lower levels, discomfort is usually reported as annoyance. ACGIH (2024) provides threshold limit values for both infra-sound and ultra-sound.

Architectural acoustics provides for the design of the acoustic environment in a space and is related to its function. In free air and open spaces, sound will reduce as the square of the distance from its source (by 6dB per doubling of distance). Indoors, the sound will be absorbed and reflected off surfaces including walls and ceilings, so sounds will linger in a room and this should be observed by the environmental ergonomist. If walls and ceiling are hard then absorbent materials such as soft furnishings and curtains, (or specialist absorbent tiles, etc.) could be recommended.

Noise, especially when not expected, can cause distraction and can influence human performance both directly (e.g. speech interference or detection of signals) by distraction and indirectly (less easy to predict). Noise can also influence mood and cause mental illness and 'irrational' behaviour influenced by attitudes to the source of the noise (e.g. neighbour annoyance).

Case studies are provided in the following chapters to demonstrate environmental ergonomics surveys of the acoustic (auditory) environment related to the health, comfort and performance of people. The role of environmental ergonomists is to provide an assessment of the environment with recommendations and that is often sufficient to meet the requirements of the client. Because of the potential complexity in some assessments, it may be necessary to refer the client to specialists in the area of acoustics and noise assessment for a more sophisticated investigation and analysis.

FURTHER READING

Kryter, K. D., 1985, *The effects of noise on man*, (2nd Ed), Academic Press, London, UK.

STANDARDS

ACGIH, 2012, TLVs and BEIs, Threshold Limit Values for chemical substances and physical agents & Biological exposure indices, *American Conference of Governmental Industrial Hygienists*, ISBN 978-1-607260-48-6.

BS EN 61672-1, 2002, Electroacoustics. Sound level meters. Specifications. www.bsi-global.com

HSE, 2005, Controlling noise at work, 2nd Ed), *The control of noise at work regulations, 2005, Health and Safety Executive guidance document*, HSE books, HMSO, UK, ISBN 0-7176-6164-4

ISO, 1999, 1990, Acoustics – Determination of noise exposure and estimation of noise-induced hearing impairment, ISO, Geneva

OSHA 1910.95, 2023, Occupational noise exposure, Occupational Safety and Health Administration; US Department of Labor, Washington, USA.

9 Case Study
An Environmental Ergonomics Survey of Noise in a Paper Mill

THE ENQUIRY

Henri is the manager of the human factors section of a large paper mill. He received a request from Mohamed who was chief medical officer in charge of the Department of Health in the company as follows.

Dear Henri

Re noise assessment of Paper mill hall 5.

I am writing to ask for your assistance in conducting a noise assessment in the above hall, with particular reference to the health of workers. The hall has had new equipment installed and I would like a preliminary indication of how the acoustic environment might affect the health of workers and what their reaction is to it. Can we meet to discuss possibilities so that we can agree on the work. The costs will be met by internal budget transfer between departments in the usual way.

Regards

Mo

THE ENVIRONMENT

Hall 5 of the papermill was familiar to both parties. Actually, Henri had conducted a heat stress survey at one of its workstations where the operator had to clean up an area where paper falls (sometimes on him) when it tears in a rapidly moving, rolling, 6 m wide, wet sheet 4 m above the ground. His team had also been involved in the design of the new control room for the process and a blueprint of the new design for Hall 5 was available.

The paper production process starts with raw materials (wood, woodchips, saw-dust, etc.) and converts it to pulp. The pulp is then cleaned and conditioned. Hall 5 receives the prepared pulp and forms it into paper when it is pressed, dried and rolled. It is then cut into the finished product and packaged. Water, steam and pulp are all involved as well as mechanical noise from rolling machines and other.

The Hall is constructed so that access is provided at heights where processes can be monitored and repairs made. Paper tears from time to time and operators restart rollers with paper after the tear. Maintenance and cleaning is also a requirement. The factories would be considered noisy and ear defenders are usually worn in some

DOI: 10.1201/9781003401964-12

areas. There have been accidents and injuries in the past so health and safety as well as worker well-being is taken seriously often involving safety culture initiatives.

THE MEETING

Henri and Debbie (who is an environmental ergonomics specialist) met with Mohamed in Hall 5, along with Alan, the Hall manager, Amelia, the workers representative (also a cleaner in Hall 5) and David from the Health and Safety section (and the union representative for the main workforce). The purpose and nature of the survey were agreed and it was emphasised that the purpose was to ensure that guidelines and legal requirements were met and that the assured health and safety of all workers was the purpose of the investigation. Alan noted that the factory operated on a shift basis and that the same activity and machines operated for 23 hours a day with one-hour shutdown at the change of the morning shift for routine maintenance and cleaning.

A report was to be provided to Mohamed and will be made available through him. The survey will involve noise measurements throughout Hall 5 to include all workplaces. Workers would also be asked for their subjective noise assessments including their views and opinions and any affects that they think noise is having on them.

Amelia and David will explain to the workers that it was in their interest to give full and honest replies to any questions and that it was the worker's own responses that were required and not that of others or a group. Suggestions would be welcome. Emphasis will be given to noise levels as objective indicators of effects on health. Subjective responses will be used as supporting information. The survey will be carried out according to the methods agreed by the Papermill Board of Trade advice notes for noise assessment. The workforce consisted of 30 papermill workers (10 per shift) – 28 identified as male and 2 as female.

AGREED AIM AND OBJECTIVES

The following aim and objectives were agreed in writing.

Aim

To conduct an environmental ergonomics assessment of noise in Hall 5 and determine actual and potential effects of the noise, on the health of workers

Objectives

1. To produce a noise map of Hall 5 with particular attention to areas where workers operate
2. To estimate daily exposure times for workers in Hall 5 and hence noise exposure
3. To obtain as supporting information, workers perception of noise levels, any experiences of effects of noise on their health and any other relevant information
4. To compare actual measured noise exposure with regulations for noise exposure and make recommendations to preserve the health of workers

ACHIEVING OBJECTIVES

Mohamed and Debbie designed the plan for the single-day investigation and Debbie carried out the survey, including noise measurements, observation and subjective recordings. Debbie wore ear defenders. It was not considered necessary to use the required 1 m square area across the floor for this exploratory survey, so 60 measurement sites were identified across the hall (50 m × 80 m) including, where possible, any workplaces including those on the upper floor and walkways. Observation of the Hall suggested that noise mainly came from transport systems for handling paper rolls, paper presses and rollers.

NOISE MAP FOR HALL 5

Objective 1: To produce a noise map of Hall 5 with particular attention to areas where workers operate

A sound-level meter and 1.5 m stand were placed at each measurement site in turn and recorded dB(lin) and dB(A) levels. The microphone was angled to provide the maximum dB(A) level at that site. The measurements across all sites were taken over the period from 9 am to 5 pm. The dB(A) values were plotted on a schematic map of the lower floor and on the upper floor with any machinery and other noise sources shown.

For both plots, contour software was used to produce lines of equal dB(A) values across the space. This provided a preliminary noise map and a prediction of the likely noise level exposure for anyone occupying any position in the space.

NOISE EXPOSURE OF WORKERS

Objective 2: To estimate daily exposure times for workers in Hall 5 and hence noise exposure

For each site, Debbie made an estimate of how long an individual would spend there each day. Additional information was provided by David as well as the workers and in particular those that occupied the site when the measurements were made. Worker profiles of noise exposure over a work shift were estimated although more accurate results will be obtained if dosemeters are used in a future study. Use of ear defenders was observed as well as any adaptive opportunities such as moving to quieter areas for a time.

The highest dB(A) value recorded was 98 dB(A) next to a rolling machine on the upper floor. A dB(lin) value of 120 dB(lin) was measured and that reflects the low-frequency component of the rollers.

EFFECTS OF NOISE ON WORKERS

Objective 3: To obtain as supporting information, worker's perception of noise levels, any experiences of effects of noise on their health and any other relevant information

A single-sheet questionnaire was given to all workers by Amelia. This was returned one week later after 23 of the 30 workers had completed the questionnaire. The results showed that workers were generally not satisfied with the acoustic environment and found it unacceptable but tolerable. They felt that ear defenders were uncomfortable in the hot steamy atmosphere and that waiting around in noisy areas was often unnecessary. Some workers felt that their hearing had deteriorated over the time they had worked in the Hall. All workers had received an annual health check including measurement of their hearing performance. Audiometry had been conducted in-house, partly for confidentiality, but some workers had in addition used a commercially available hearing test (widely available through 'opticians').

Jean was a machine operator in the factory who had indicated that her identity be kept confidential but that her responses could be used as general information. She had found the ear defenders difficult to wear but did so. She had had difficulty in hearing speech and her partner had become frustrated with cumbersome conversation.

She attended the local 'optician' for a hearing test. She sat in a booth and single-tone sounds had been played to her through earphones. Sometimes there was no sound present (monitors guessing). She had to press a button when she heard a sound and release it when she heard no sound. This is called the up and down method where the absolute threshold of hearing is determined by a single frequency sound, well above threshold, being played through the earphones in pulses and gradually reducing in level. When the level was established where Jean did not press the button, the level of the pulses increased until the button was pressed again. By alternating up and down in level, the threshold value for that frequency (the lowest sound that could be heard) was recorded and the threshold for the next (third octave) frequency was determined. This was repeated for both ears.

An audiogram provides a comparison of hearing performance (threshold) with that for a normal healthy ear in the population. It is plotted on a spectrum with a 0 dB straight line indicating no 'drop' (difference) between a 'normal' threshold and a tested ear. The threshold for the person is then plotted on the audiogram and shows where hearing performance is lower than normal.

For Jean, there was a small drop at higher frequencies which is associated with age but significantly there was a 6 dB drop at a frequency of 4kz. This spread across the frequency range into frequencies associated with speech (0.25kHz to 4kHz) and is a clear indication of industrial hearing loss.

The cochlear of the ear receives sound as mechanical vibrations and has hairs sensitive to different frequencies. High-level industrial noise can damage the hairs specific to detecting 4kHz and this can spread to adjacent hairs with prolonged exposure (years).

Three workers reported minor hearing loss and five reported ringing in the ears in the evening. There was sufficient comment to recommend a review of the health screening and hearing protection procedures for the Hall and training of staff in behaviours to minimise noise exposure.

Ear defenders require further investigation for thermal comfort and wearability. It is interesting that the performance of ear defenders is determined in a similar way to that of audiometry (more controlled and a group of healthy people measured with and without the ear defenders in an anechoic chamber at the centre of speakers located at the apexes of a specified trapezoid space). Measurement of the efficacy of ear defenders may be useful in the actual noise conditions of the Hall using probe microphones.

NOISE ASSESSMENT

Objective 4: To compare actual measured noise exposure with regulations for noise exposure and make recommendations to preserve the health of workers

Health and Safety regulations (HSE, 2005a) state that it is a legal requirement to ensure that workers are not exposed to greater than 85 dB(A) Leq for an 8-hour shift. So 88 dB(A) Leq; 91 dB(A) Leq; 94 db(A) Leq; and 97 dB(A) Leq for exposures of 4 hours; 2 Hours; 1 hour and 30 minutes, respectively.

Although dose and Leq values were not taken, the noise map provided in Figure 9.1, along with reported and observed worker behaviour and predicted exposures (times and places) indicate that a safe strategy would be to implement procedures to ensure that all people who occupy Hall 5 wear hearing protection that ensures an exposure of at least 10 dB(A) reduction to noise levels in the ear.

RECOMMENDATIONS

1. General
   a. A detailed (1 m^2 grid) noise map should be produced for Hall 5.
   b. Actual dB(A) Leq dose values should be measured on workers especially those with long-term exposure (years) in hall 5.
   c. Health screening procedures should be reviewed and a new procedure implemented if necessary.

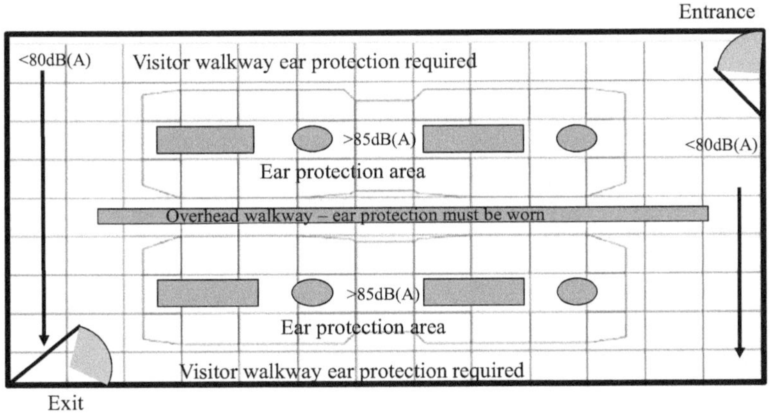

FIGURE 9.1 Noise map of paper mill Hall 5 showing noise levels on walkways and when working near machines.

2. Attempts to reduce noise at source and worker exposure
   a. Noise reduction experts should be employed to explore possibilities for reduction of noise at source.
   b. Work design should be reviewed to reduce noise exposure while maintaining performance and productivity. Quiet (cool) areas for breaks should be implemented and work/rest regimes or alternating work to less noisy areas to reduce dose should be investigated.
   c. Specific attention should be given to the location of walkways through Hall 5 to ensure minimum exposure for both workers and visitors.
   d. Methods for ensuring ear protection is worn, in an effective way, should be investigated. Reasons for reluctance to wear ear defenders should be addressed and rectified.

3. Health monitoring and screening
   a. Existing screening and monitoring methods should be reviewed including initial, and regular monitoring, audiometry and a system of reporting for workers.

4. Health and safety issues
   a. Review methods to ensure that no worker is exposed to greater than 85 dB(A) Leq and hence that legal requirements are met.
   b. Display appropriate signs, reminders and warnings across Hall 5 regarding noise exposure. Include a noise map at entrances to Hall 5 and a briefing regime for visitors.

Notes

1. Recommendations are made above as findings of the environmental survey for use by the 'in-house' Department of Health and Health and Safety Department, who will have a more profound view and expertise in their own areas.
2. A useful method of determining the effects of noise on human performance and productivity is the RDC (Regulation; Distraction; Capacity) model (Parsons, 2021). This involves time off task so that the requirement to not exceed legal limits in noise Regulations can be met; any time off task due to noise Distraction, including annoyance and discomfort; and loss in Capacity to carry out the tasks. This will involve task analysis and job design so for example if speech communication is essential then the effects of noise can be determined and reduced capacity due to noise, estimated.

OUTCOMES

The report was sent to Mohamed for distribution. The health department implemented revised procedures for determining the hearing performance of workers. It involved screening to ensure those unsuitable for work in the environment were prevented from doing so and to establish a baseline for new employees and current position for existing employees.

A six-month review was established and audiometry was sub-contracted out to a commercial 'opticians' to provide convenience and independence. Workers were encouraged to report any affects they thought that the environment was causing to their health both at work and at home through a focus group method that did not raise alarm. A discussion and suggestion group was established with significant management representation.

An unexpected consequence was that the environmental ergonomics part of the Human Factors section was transferred to be the responsibility of the Health and Safety Department who had received the report and said that they will act upon it.

Jean sued the company and received substantial damages, including an amount for a bespoke hearing aid system that specifically compensated for her hearing damage.

FURTHER READING

Kryter, K. D., 1994, *The handbook of hearing and the effects of noise*, Academic Press, San Diego, CA.

STANDARDS

BS EN 61672-1, 2002, Electroacoustics. Sound level meters. Specifications. www.bsi-global.com

HSE, 2005a, Controlling noise at work (2nd ED), *The control of noise at work regulations, 2005, Health and Safety Executive guidance document*, HSE Books, HMSO, UK, ISBN 0-7176-6164-4

HSE, 2005b, *Protect your hearing or lose it, Pack of pocket cards*, HSE Books, ISBN 0-7176-2139-1

ISO, 1999, 1990, *Acoustics – Determination of noise exposure and estimation of noise-induced hearing impairment*, ISO, Geneva

ISO 28002, 2012, *Ergonomics of the physical environment -The assessment of environments by means of an environmental survey involving physical measurements of the environment and subjective responses of people*. ISO, Geneva.

10 The Effects of Noise on Annoyance, Communication and Task Performance

NOISE ANNOYANCE

Noise can cause discomfort often expressed as annoyance. Annoyance can be measured using subjective scales and predicted from observation of behaviour and from physical measures, particularly dB(A) levels. All three are usually involved in an environmental ergonomics survey of the acoustic environment.

Annoyance is not directly related to noise level and individual disposition, preference, expectation and attitude are important. Reducing noise levels may not reduce annoyance. The same noise levels in a quiet room may cause different levels of annoyance than in a busy office or outdoors in a railway station. Some engine noise on a ship or aeroplane may be reassuring. Short bursts of noise can be startling and people may adapt to background noise but are sensitive to change.

Indices have been developed for particular contexts. Haslegrave (2015) describes a selection and refers to Beranek (1971), Ollerhead (1973), Schultz (1978), Fidell et al. (1979), Kryter (1985), Loeb (1986) and ISO 1996-1 (2003) for detailed information. Collectively they have produced the noy unit based upon equal annoyance curves – the noise rating index (NR); the noise number index (NNI); the traffic noise index (TNI); the noise pollution index (NPI); the day-night average sound level Ldn; preferred noise criteria (PNC) curves; and others. They cover contexts indoors, outdoors, in different types of spaces (rooms, transport, etc.) and for different types of noise characteristics.

Interestingly there has been little concentration on individual characteristics (including hearing performance) and population diversity (age, gender, ethnicity, etc.). The indices often use dB(A) and Leq measures in their formulae as well as frequency analysis and additional contextual factors. For the environmental ergonomist, measuring dB(A) levels across a space where people are exposed to the noise, is a good starting point and often sufficient to make recommendations or as a basis for further investigation.

NOISE AND PERFORMANCE

The concept of performance has no meaning in isolation and must be considered and defined in terms of a task being performed. It is measured by the level of an agreed

DOI: 10.1201/9781003401964-13

attribute or attributes. Examples are number of bricks used per hour when building a wall, marks in an examination or satisfaction level in a feedback questionnaire. Productivity is usually related to the goals of a group or organisation and often integrates a number of tasks to achieve that overall goal (e.g. profit or factory throughput). To consider the effects of an environment on performance and productivity, therefore, the terms must be clearly defined along with their method of measurement.

ISO TR 23454-1 (2024) provides a generic conceptual model from which the effects of the environment on human performance can be constructed. It considers performance as a multiple of factors related to the influence of the environment on performance in terms of Health and Safety; Distraction; and Capacity. The HSDC (or RDC where R is for Regulations) method uses the percentage of time allowed for performing the task (i.e. the exposure time allowed after the implementation of environmental limits or Regulations); multiplied by the percentage of time on the task after taking off any time not performing the task due to Distraction; multiplied by the percentage of Capacity remaining after the influence of the environment when compared with optimum conditions (e.g. comfort).

Suppose the noise level limit in a factory was 85 dB(A) for an 8-hour Leq and the level recorded was 88 dB(A). Then, the regulations would permit work in that environment for only 50% of the time. If workers were distracted for 10% of the time during that period, then we have 90% of that time on the task. Finally, if the noise reduced capacity by 10%, then workers will work at 90% capacity. Overall performance is then given as $R \times D \times C = 0.5 \times 0.9 \times 0.9 = 0.405 = 40.5\%$ of that of work in optimum conditions. If noise levels meet regulations and there was no loss in capacity but the same level of distraction, then $R \times D \times C = 1 \times 0.9 \times 1 = 0.9$ or 90% of that of work in optimum conditions (ISO TR 23454-1 (2024).

The effects on performance due to the regulations are direct. If noise levels are above limits, then it can be assumed that it is time off the task. Other tasks in other environments are possible (during that 'downtime'), in work design, but that is not considered here. For hearing loss, 85 dB(A) Leq is often used as a limit for industrial work so doubling the noise level to 88 dB(A) Leq would reduce performance and limit exposure time by 50% (91dB(A) Leq limits exposure time to 25% and so on). Where noise limits are not exceeded, exposure time will be 100% of work time.

Noise can cause distraction, particularly when not expected. The extent to which an individual will be distracted by noise will depend upon 'startle, novelty, curiosity and alarm' as well as psychological factors such as attitude to the source of the noise, the degree of annoyance and dissatisfaction with the noise and motivation to complete the task. It includes attending to the noise in an effort to reduce it, hence taking time off task to do so and maybe disturbing and distracting others in the process. Wearing ear protection and behavioural responses may provide less exposure to noise but will also cause distraction from the task.

The effects of noise on Capacity to carry out tasks can be considered as those that directly affect hearing performance and general effects. Hearing performance can be in terms of speech interference both as a person as receiver or the person's voice received by other people. It can also include detection and interpretation of information, signals and alarms. A general principle for all environmental components could be that performance will be affected when unwanted stimuli compete for the same resources (e.g. noise, speech and the auditory system).

GENERAL EFFECTS OF NOISE ON PERFORMANCE

A general effect of noise on performance is distraction (considered above). General affects on the capacity to carry out tasks are not easy to predict and noise may improve performance (increase capacity), have no affect or reduce capacity. For the same noise environment, all three outcomes have been found in laboratory experiments, implying that there are confounding factors related to individual disposition and context.

A popular explanation is that a person's level of arousal (wakefulness, readiness to respond, base energy level, etc.) influences performance. If a person is performing a boring monotonous non-stimulating task, then high levels of noise (or other stimuli) increase arousal towards an optimum level and performance is increased by noise. If the task or job is stressful and causes a high level of arousal then noise can cause a person to be over-aroused and performance will decrease. This sometimes leads to proposals to introduce desirable sounds (not noise), or 'white noise' to mask other noise, in an attempt to improve performance.

SPEECH INTELLIGIBILITY

Speech is a complex sound that varies in frequency, level and phase as well as temporal variation all formulated through evolution, culture and cognitive and physiological development. It can be generated, received and understood by people. Hearing loss (and other damage to the auditory system) in the frequency range for speech (0.25 to 4 kHz) reduces intelligibility and noise (in general background noise) can mask speech (where one sound dominates another).

People can interpret speech even when a substantial part of the sound is masked by noise (particularly if the speech is expected, facial and mouth shapes are seen, or in a context familiar to the receiver). Once a sound has been detected, the brain can focus on the 'signal' and enhance detection (irritating when the signal is unwanted).

Speech intelligibility in noise can be measured by exposing people to nonsense syllables (simple phonetically balanced words with no meaning (e.g., in English – boh, tay, etc.) in background noise (e.g. white noise containing all frequencies at the same level, or specific background noise from the environment and so on) or actual speech (or selected words and so on) of interest. An individual listens to the words or phrases and reports what they hear. Interference and intelligibility can then be quantified as number correct or with more sophisticated measures using confusion matrices.

Haslegrave (2015) describes three methods for predicting speech intelligibility using speech; background noise; and task (e.g. distance from speaker) characteristics (See ISO 9921, 2003). These are the Speech Intelligibility Index; the Speech Interference Level (SIL); and a direct method for estimating SIL. All involve dB(A) levels and frequency characteristics of the background noise.

The Speech Intelligibility Index involves the speech characteristics and is an integral of the signal-to-noise ratio values in third-octave bands for the listener, weighted according to speech intelligibility importance. Additional contextual factors (reverberation; background noise interruption, etc.) can be added. The index value can be interpreted as a prediction ranging from 'difficulty likely' to 'good speech communication possible'.

The SIL does not involve actual speech and predicts speech interference from the mean of sound pressure levels (SPL) in four-octave bands. The SIL is interpreted for normal, raised and shouting speech levels for a range of distances between speaker and listener. Loeb (1986) provides an estimate of the SIL as 9 or 10 dB below the SPL, and Webster (1979) provides a chart relating dB(A) levels along with voice effort and speaker-hearer distance to quality of communication.

In an initial environmental survey of the acoustic environment, it is usual to measure only dB(A) levels in the first instance, and the possibility and consequences of speech interference will be estimated directly from those and from observation and subjective measures.

An interesting phenomenon is sub-conscious competing to hear and be heard in a crowded room, for example. 'Acoustic spiralling' occurs as noise intensity spirals upwards seemingly out of control. Any school teacher will know that if they allow background chatter, it will build to a crescendo as voices of individuals spiral in level to outdo background chatter but rendering their efforts ineffective as others do the same.

WARNING, ALARM AND INFORMATION SYSTEMS

Warning or alarm systems are usually either auditory, visual or both and require immediate detection and interpretation for action. They are designed to cause distraction. Information systems can be via acknowledged code, speech or generated speech and need to provide a clearly detected and understood message. Five (low frequency) blasts from a vessel at sea is a standardised signal to another vessel that it should make its intensions clear (e.g. to facilitate avoidance before a collision takes place). The 'bing-bong' sound in a railway station alerts travellers that a message is about to be transmitted and so on.

The principle of design is that the signal should be above the background noise and any other sounds. More specifically, the level and frequency of the signal should be sufficiently different from the background noise that it can easily be detected and recognised (without causing pain, startle or over-reaction or hearing damage). It is also relevant that people are particularly adept at detecting change and contrast in their environment.

In summary, the design of environments that allow speech intelligibility as well as detection of signals, alarms and messages on information systems can be facilitated from measurements and analysis or assumptions about the characteristics of background noise and the characteristics of the sound to be detected. The procedure involves the level and frequency characteristics of the signal and the background noise.

The environmental ergonomics survey would not normally make these detailed measurements and calculations. A first and often sufficient method is to use subjective methods to determine ease of communication or detection and to construct simple listening tests. The intensity of sound deteriorates as the inverse square of the distance from the source (level at listener = level at source/(distance)^2) and is influenced by objects and barriers. Observation of communication by the investigator is also of use. User trials of alarms are essential.

Chapter 11 demonstrates the principles of the environmental ergonomics noise survey for consideration of noise in a large open-plan office used as a call centre.

FURTHER READING

Haslegrave, C. M., 2015, Auditory environment and noise assessment, In Wilson J R and Sharples S., Evaluation of human work (4th Ed), CRC Press, Taylor & Francis Group, Boca Ratan, New York, Oxford, ISBN 978-1-4665-5961-5, Chapter 26, pp 705–724.

STANDARDS

ISO 1996-1, 2003, Acoustics - Description, measurement and assessment of environmental noise: Part 1-Basic quantities and assessment procedures. ISO, Geneva.

ISO 9921, 2003, Ergonomics – Assessment of speech communication, ISO, Geneva

ISO TR 3352, 1974, Acoustics - Assessment of noise with respect to its effects on the intelligibility of speech, ISO, Geneva

11 Case Study
Environmental Ergonomics Survey of noise in a Call Centre

THE ENQUIRY

Haruto is the manager of an open plan office that is an international call centre and part of a multi-national insurance company. He contacted the human factors and ergonomics department of the University of Fukoyo for advice as follows.

Dear Sir or Madam

I am the manager of a large open plan office. I am writing to ask for your advice on how to ensure that the acoustic environment in the office is appropriate to ensure the well-being and productivity of employees. Operators work in cubicles in the office, wear earphones, mostly have telephone conversations with clients and potential clients and work on computers. Operator performance is directly related to the productivity of the office and as well as quality of life I believe that maintaining worker satisfaction is important and obviously I would wish to ensure that there is no damage to health.

I hope that you will be able to help.

Yours sincerely

Haruto

Manager, World-wide communications

THE MEETING

Paul, the Head of the Human Factors and Ergonomics Department at Fukoyo University, passed the enquiry on to Minato, Head of Environmental Ergonomics. Minato and Akari (Head of Acoustics) met with Haruto at the office and after a review of the work areas and tasks, the following aim and objectives were agreed for an environmental survey of the acoustic environment.

AGREED AIM AND OBJECTIVES

AIM

To conduct an environmental ergonomics survey of the acoustic environment in the call centre office, to assess the effects of noise on the health, comfort and performance of the occupants.

DOI: 10.1201/9781003401964-14

Objectives

1. To measure the acoustic environment in the office, with particular reference to the noise exposure of occupants, and to produce a noise map
2. To determine the workers' perceived effects of noise on their health, comfort and performance
3. To compare the acoustic environment with guidelines and regulations for call centres
4. To provide an overall assessment of the office and make recommendations to enhance the health, comfort and performance of operators
5. To provide a report to Haruto within 1 month of the survey with an executive summary and describing the survey and its findings

Haruto emphasised that the office was international and operated a shift pattern over 24 hours. There were 100 operators in four shifts (morning, 6 am to 12; afternoon, 12 to 6 pm; evening, 6 pm to 12; and night, 12 to 6 am) of 25 operators, all of whom identified as female. The work was intense and time off task involved in the survey should be minimum.

Minato explained that the noise measurements and observations will be conducted once for each shift at a typical period and workload. Operators should be informed of the timing and that the survey was for their benefit. The noise measurements and observations should not interfere with workers and questionnaires should be completed as close to the measurement time as possible but at the convenience of the operators. Operators should respond with their own views and not those of others. Operator location will be required but individual identity will be kept confidential by Akari.

Haruto agreed to send an email to all workers informing them of the survey and asking for cooperation in completing the questionnaires. Haruto will not be involved in the survey but will carry out his normal duties on the days of the survey. Akari will be the sole person conducting the survey and she will distribute questionnaires at the beginning of the measurements and collect the completed questionnaires at the end of the shift. She will attend on four days, one for each shift.

ACHIEVING OBJECTIVES

Measurement and Presentation of Noise Exposure

Objective 1: To measure the acoustic environment in the office, with particular reference to the noise exposure of occupants, and to produce a noise map

A blueprint of the office was obtained with furniture, equipment and workstations identified. The office was 50 m × 30 m and contained 30 workstations consisting of a computer workstation and telephone and headphone equipment. Locations of measurements were identified at the workstations and at the centre of 5 m^2 grids across the whole office. The noise measurements began after one hour from the beginning of each 6-hour shift. A sound-level meter (B&K type 2245) was used to provide dB(A) and dB(lin) levels near the position of the ear of the operator and representative values over one minute were recorded.

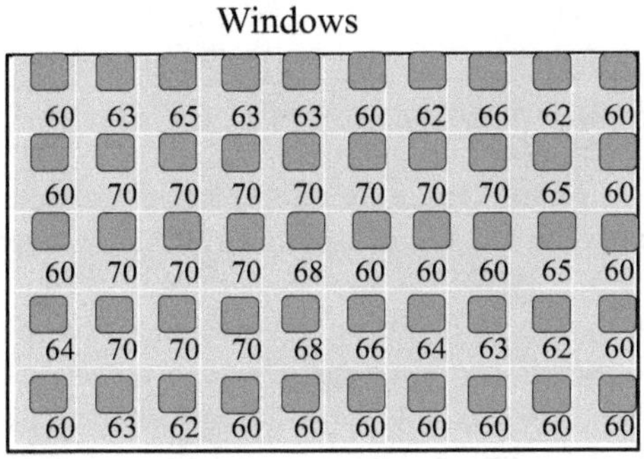

FIGURE 11.1 dB(A) noise levels in the call centre, at each workplace.

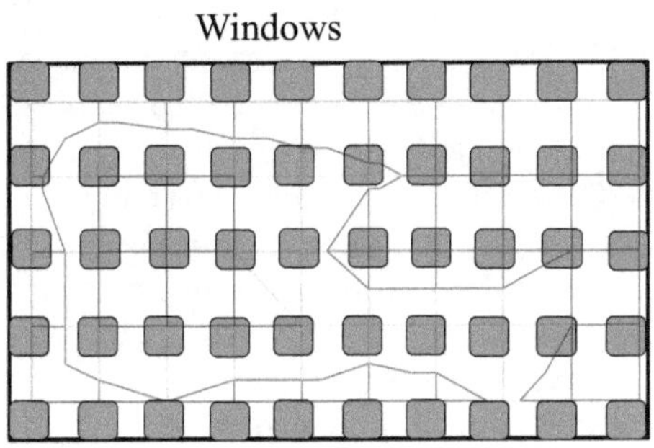

FIGURE 11.2 Noise map in the call centre.

Each dB(A) value was plotted on a blueprint of the office for all four shifts. A second presentation was a noise map of the office using the dB(A) values and contour map software (Figures 11.1 and 11.2). It can be seen that the noise levels in the office ranged from 60 to 70 dB(A).

Observations suggested that this was all 'chatter' and that background noise was therefore chatter and was not increased by traffic or other outside noise. The noise levels were consistent across all shifts demonstrating the international range of work roughly focussing on Europe, the USA and Asia but with only the English language.

Earphones provided a level of privacy and would be expected to preserve effective communication. There seemed to be no evidence of a crescendo effect (acoustic spiralling) where voices spiralled upwards in level as operators tried to make themselves heard. A consistent level of noise was present throughout the shift and operators worked continuously apart from a 10-minute break every 2 hours between calls.

EFFECTS OF NOISE ON WORKERS

Objective 2: To determine the workers' perceived effects of noise on their health, comfort and performance

Akari provided Haruto with interactive software consisting of a questionnaire for operators to complete concerning how they felt about the effects of the noise, at their workstation in the office, on their health, comfort and performance. Haruto attached the questionnaire link to his email and asked operators to complete it fully during quiet periods over the shift and that if necessary they could take up to 15 minutes off task in order to complete the questionnaire. The completed questionnaires were sent directly to Akari.

The questionnaire consisted of rating scales and open-ended questions. There were three similar sections concerning the effects of noise on THEIR OWN health, comfort and performance, respectively. The first request was – Please rate on the following scales and provide further information on how YOU feel noise affects YOU at work. The scales were based upon ISO 10551 (2019) beginning with 'I find the noise in MY working environment to be…': 1. Not annoying; 2. Slightly annoying; 3. Annoying; 4. Very annoying; a personal preference scale: 1. No change; 2. Slightly quieter; 3. Quieter; 4. Much quieter and forced choice scales of Acceptable/Unacceptable; and Satisfied/Dissatisfied. Open-ended sections were added to allow for further and explanatory comments. A scale of distraction prefaced 'please rate on the following scale how Distracting you find the noise in your environment': 1. Not distracting; 2. Slightly distracting; 3. Distracting; 4. Very distracting. Sections prefaced by 'If you think that the noise in your environment is affecting your health then please provide details' as well as 'If you think that the noise in your environment is affecting your ability to carry out the work then please provide details'. A final section requested any other comments relevant to the noise YOU experience at work, including causes and suggestions for improvement.

An observation checklist was completed by Akari that included general impression of the acoustic environment; behaviour of the operators; and impression of interference with speech communication (not through earphones), including making conversation with others in the office where available; and specific problems that were apparent during the survey.

The results for all three shifts were similar although noise on the night-time shift (local time 12.0 pm to 6.0 am) was less annoying, corresponding with the slightly lower levels measured on that shift. People were generally: between 'Not annoyed' and 'Slightly annoyed' but two of the operators were very annoyed and commented that some of the operators were unnecessarily loud in their work; would prefer it to be quieter but commented that they understood it was the nature of the work. For the same reason, 90 of the 100 workers were satisfied with the acoustic environment and found it acceptable.

Operators were generally not distracted apart from some (12 of the 100) distracted by an old, large printer in the corner of the office, by others' activity when they were communicating and some operators with unusually loud 'telephone' voices. They did not feel that this affected their performance as they were highly motivated to provide a high-quality interaction with customers and receive good feedback.

Comments included often feeling thirsty (30 out of 100) and headaches and 2 people had effects on vision. The main comment on health was after work when ringing in the ears (tinnitus) was disturbing mainly when trying to sleep (15 of 100 workers) and ear infections (10 of 100 workers) maybe caused by lack of earphone hygiene. No significant differences were found across shifts.

The results of the subjective and behavioural assessments suggest that operators were generally satisfied with the acoustic environment, found it acceptable and did not regard it as interfering with their performance. It is recommended that: telephone training should include voice level along with communication skills; that the printer should be moved or replaced; and that health screening and training should include questions concerning problems after work, with an appropriate action plan.

ASSESSMENT OF NOISE

Objective 3: To compare the acoustic environment with guidelines and regulations for call centres

Noise regulations for work (e.g. HSE, 2005a), are provided as dB(A) Leq values and typical limits are for lower exposure values: daily or weekly exposure of 80 dB (A) and peak sound pressure of 135 dB(C) and for upper exposure values: daily or weekly exposure of 85 dB(A) and peak sound pressure of 137 dB(C). The maximum dB(A) values measured over-all shifts during the survey were 70 dB(A) and no very loud (peak) sounds were evident. It can be concluded therefore that the environmental noise was within regulations. It is also reasonable to conclude that sound levels at the ear through earphones did not exceed regulations.

Although not a regulation regarding health, offices and call centres should aim for a maximum general noise level of below 60 dB(A) and it should be noted that the call centre investigated has noise levels above that guideline.

Objectives 4 and 5 are met in the overall assessment provided in the report described below and sent to Haruto within one month of the completion of the survey.

REPORT: AN ENVIRONMENTAL ERGONOMICS ASSESSMENT OF THE ACOUSTIC ENVIRONMENT OF A CALL CENTRE

Executive Summary

An environmental ergonomics survey of the acoustic environment was carried out at the call centre of the International Communications office on 5, 10, 15 and 20 November 2024 (morning, afternoon, evening, and night shifts, respectively). The aim was to assess the effects of noise on the health, comfort and performance of the occupants. Noise levels were measured across the office and subjective and behavioural measures were taken using a questionnaire and an observation checklist.

Similar results were found for all four shifts with average dB(A) levels of 65, 70, 67 and 63 dB(A) for morning, afternoon, evening and night shifts, respectively. These are within the regulations for noise exposure that prescribe lower limit values of 80 dB(A), with no individual measurement exceeding those limits, no sudden very loud (peak) noises and no significant contribution to noise levels from outside of the office. It is recommended that a target value of around 60 dB(A) should be set. Although not required by regulations, it is more typical of call centre environments and may be beneficial to workers.

Background noise was mainly chatter and operators were highly motivated and concentrated on the task in hand. They generally found the noise environment to be acceptable and were satisfied within the context of the work that was carried out. They would prefer it to be quieter but understood the nature of the work and generally did not find the noise to be annoying nor distracting.

A printer caused some annoyance and distraction as well as two operators that were identified as having unacceptably loud telephone voices. Damage to hearing would not be expected and was not reported in the survey. There was, however, an indication that tinnitus (ringing in the ears) occurred after work and that ear infections were prevalent, possibly caused by earphones. It is recommended that the printer be relocated or replaced and that health screening and health affects be further investigated. Operator training should include voice level when addressing communication skills.

SUMMARY OF REPORT CONTENTS

Objectives

1. To measure the acoustic environment in the office, with particular reference to the noise exposure of occupants, and to produce a noise map
2. A noise map was provided (Figures 11.1 and 11.2) on a blueprint of the call centre for each of the shifts, presenting dB(A) levels with date and time recorded and by whom. Mean, range and variation statistics were provided.
3. To determine the workers' perceived effects of noise on their health, comfort and performance
4. An annoyance map was provided on the blueprint of the call centre with median and range of values. A description of ratings and comments was also provided. It was concluded that generally there were few effects on health and performance although specific areas required further investigation.
5. To compare the acoustic environment with guidelines and regulations for call centres
6. Details were provided including the conclusion that all noise levels were below the regulation 80 dB(A) level and hence met existing standards for health and safety.
7. To provide an overall assessment of the call centre and make recommendations to enhance the health, comfort and performance of operators.

Recommendations reported were that: further investigation is required on any affects after work on hearing; specific causes of annoyance should be addressed; and that although performance did not appear to be affected by noise, any noise reduction may be beneficial.

Additional Information

A useful description of noise in call centres is provided by KRISP(2024). They note that distracting sounds in call centres include people chattering – telephones ringing; keyboards clacking; and fax machines running. Modern computer workstations would eliminate some of those distractions. Noise reduction devices are available, however, 20 suggestions for improving the acoustic environment are made as follows:

1. Increase space between workstations.
2. Use noise cancelation systems.
3. Install acoustic panels.
4. Use low noise-affected headsets.
5. Use sound masking (e.g. white noise).
6. Use partitions between workstations.
7. Use silent keyboards.
8. Use software for ringtone (through earphones).
9. Use flooring to avoid footfall.
10. Train agents for voice volume.
11. Use several smaller offices rather than one big one.
12. Use alternatives to telephone calls (e.g. chat lines).
13. Use meeting rooms separate from the call centre.
14. Consider letting agents work from home.
15. Move noise makers (printers, refrigerators, etc.) to a different room.
16. Use virtual internal meetings and communication systems.
17. Use indoor plants to disperse sound (but avoid hay fever).
18. Regulate how people move across the office.
19. Observe legal limits and requirements (assess hazards, reduce exposure, provide training and hearing protection).
20. Encourage operators to listen to themselves, leading to personal noise reduction.

The report (and invoice) marked 'commercial in confidence' was sent to Haruto within 1 month of the survey with an executive summary and describing the survey and its findings.

OUTCOMES

The report was well received as meeting objectives but no other feedback or clarification was sought. It is assumed the recommendations were considered and acted upon. Workers will have expectations, so a lack of response can cause dissatisfaction. A moral dilemma is that there are possible affects of the sound levels in the call centre on health. This is a general dilemma for the consultant, not specific to noise surveys, and requires consideration.

FURTHER READING

Krisp, A., 2024, *Noise cancellation and artificial intelligence. krisp.ai/blog/call-center-background-noise/*

Kryter, K. D., 1985, The effects of noise on man, (2nd Edition), Academic Press, London, UK.

STANDARDS

ISO 28802, 2012, Ergonomics of the physical environment – Assessment of environments by means of an environmental survey involving physical measurements of the environment and subjective responses of people, ISO, Geneva.

Part IV

Environmental Ergonomics and Vibration

The effects of vibration on people are particularly important in vehicles and when holding vibrating objects. Low-frequency vibration causes motion sickness and higher frequency vibration causes discomfort, loss in performance and effects on health as levels increase. Vibration levels close to threshold values can cause alarm and disturbance in buildings. Hand-transmitted vibration can cause damage to the hands, wrists and arms. Part 4 considers whole-body vibration in three chapters (Chapter 12, 13 and 14) and hand-transmitted vibration in two chapters (Chapters 15 and 16). It provides principles of assessment as well as case studies concerned with vehicle ride, office vibration and working in a dockyard.

From the late 1960s, Mike Griffin and his team at the Institute of Sound and Vibration Research, Southampton, United Kingdom, turned human response to vibration from a specialist underdeveloped subject to an accessible mature developed subject, through research orientated towards application. Together with international researchers throughout the world, numerous international standards have been produced for use in application. Developments in technology have played a major role such that the assessment of vibration can be an integral part of environmental ergonomics. Methods agreed nationally and internationally are presented in Part 4.

In Chapter 12, mechanical vibration is described in terms of magnitude (level), wavelength and frequency, and when it is in contact with the whole body, it causes it to vibrate. Units derived from the time variation of vibration include the peak; peak to peak and root mean square (rms) values, and the internationally recognised measure is acceleration in ms^{-2}. As well as level and frequency, vibration inputs to the

DOI: 10.1201/9781003401964-15

body simultaneously vary in six axes; fore-and aft; side-to-side; vertical; roll; pitch; and yaw and at various inputs. These are described for a seated subject as at the feet; ischial tuberosities; and back. Frequency weighting functions (wb, wc, wd, we, and so on) are presented that are used to derive values related to human response (w_{rms}) for each axis and input of vibration. (Figure P4). When combined they provide an integrated effect over all axes and inputs of vibration. The vibration dose value (VDV) is described as a method for considering the effects of vibration exposure time.

In Chapter 13, Gordon was asked to obtain credible publicity material for his company that had moved from manufacturing petrol to electric cars. His task was to obtain evidence that the electric car had a superior ride to that of the petrol-driven car. He contacted the internationally respected University of Solenti and an environmental ergonomist carried out a field trial to compare the vibration ride quality in, an otherwise identical, petrol-driven and electric cars.

Four male and four female members of the general public were driven around public roads, each in both cars. Acceleration levels were measured at the feet, seat and back of the seated passengers. Subjective scales were completed by the passengers and at the end of the journey, a car preference was provided in terms of vibration ride. Vibration indices were calculated using appropriate weighting functions and integration methods. After consideration of all findings, it was concluded that the

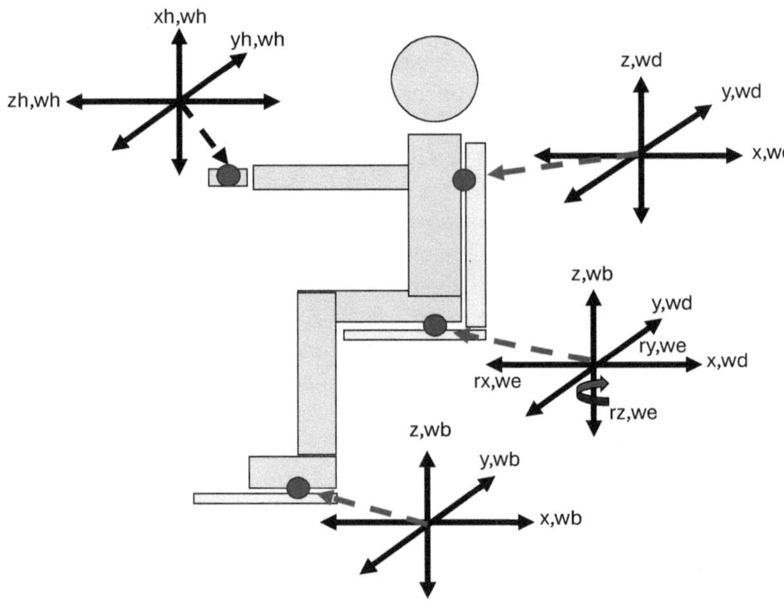

FIGURE P4 Schematic representation of vibration inputs to a (seated) person showing coordinate systems at the feet, ischial tuberocities, back and hand. For inputs at the seat, vibration axes are: x, fore-and-aft; y, side-to-side; z, vertical; rx, roll; ry, pitch; and rz, yaw. Corresponding frequency weighting functions: wb; wc; wd; we; and wh used in the prediction of human response, are shown. Suffix h indicates hand vibration.

electrically driven car provided a significantly more comfortable vibration ride than the car fuelled by petrol.

Chapter 14 considers vibration in an office. Maria is the manager of an office above an underground railway station. She contacted Martin, a government-employed Environmental Ergonomist, to reassure workers that vibration levels were within regulations. Floor and seat acceleration was measured at workstations across the office and over the working day. Subjective measures were also taken from workers on a standard questionnaire. It was found that vibration levels were above vibration detection thresholds when trains passed through the station but that when taken over the whole day, they were below levels that would cause adverse comments according to national and international standards. Workers generally indicated that they detected the vibration but were not concerned. They were re-assured about the vibration levels but the office environment in general had become a topic for discussion.

Hand-transmitted vibration (Chapter 15) from vibrating tools can cause damage to blood vessels; nerves; muscles; bones; and joints of the hands, wrists and arms. The multitude of factors involved make an assessment of the vibration and prediction of effects particularly difficult. Health screening, diagnosis and subjective methods can be used to quantify effects, although cause may be difficult to identify. Measurement at the hand or on the tool handle involves accelerometers in three orthogonal axes with respect to the hand or tool handle (fore-and-aft; side-to-side; vertical) (Figure P4). Tentative-weighting functions (wh) are defined in international standards and weighted root-mean-square acceleration values (whrms ms^{-2}) provide some validity and are termed A(8) for an exposure over an 8-hour day. An A(8) value of 2.0 ms^{-2} is tentatively considered as a limit below which symptoms of hand-arm vibration syndrome are rare.

Chapter 16 presents a case study of the assessment of vibration at work in a dockyard involving vibration tools. Noah had been asked to make a risk assessment of work in a dockyard where some of the workers use metal grinders. At Noah's request, Hudson, from the Naval Medical Research Centre, conducted an environmental ergonomics Assessment of the effects of the hand-transmitted vibration. Three of the 20 workers conducted a one-minute, grinding task on a ship's hull, while measurements were made with appropriately positioned accelerometers. All 20 metal grinder workers completed a subjective questionnaire that was also completed by a control group of office workers.

The worst-case exposure was a weighted value of 21.3 ms^{-2} (worker B, right hand), and when compared with regulations of 2.5 ms^{-2} action limit and 5.0 ms^{-2} for 8 hours the action limit would be reached after 7 minutes and the exposure limit after 26 minutes. That is 7 and 26 operations respectively. Subjective assessments implied that 15 of the 20 workers had symptoms of damage to the hands and that three probably had vibration white finger and two workers showed symptoms of carpel tunnel syndrome. The control group showed no significant effects in comparison. On receiving the report, Noah instructed that workers should be exposed to no more than 26 exposures per day, a system of health checks shall be implemented immediately and a more comprehensive survey should be carried out.

12 Human Response to Whole-Body Vibration

VIBRATION

Mechanical vibration is the oscillation of solid material and when it is in contact with the human body, it causes the person to vibrate with possible consequences for the health, comfort and performance of the person. The material oscillates as a transverse wave and hence involves a displacement of material with time about a central position.

The wavelength is the distance from one point to another corresponding point when the wave repeats itself and can be represented as the time to complete one wave or cycle (s). The frequency is the number of waves completed in 1 second (1 Hz = 1 cycle per second) and the level is the magnitude of the displacement (m), and as it is time-varying, it can also be considered in terms of velocity (ms^{-1}), acceleration (ms^{-2}) and more. When considering human response to vibration, the magnitude (level) of vibration is usually expressed as acceleration and is measured using accelerometers.

The average magnitude of vibration is usually expressed as root mean square (rms) acceleration, the square root of the integral of acceleration squared with respect to time divided by the time under consideration (in digital form – square all values taken at equal time intervals, add them, divide by the number of values and take the square root), but other parameters are peak-to-peak (maximum minus minimum value over a time period); peak (maximum deviation from a central position) and more. The root mean quad (rmq) value is similar to the root mean square value but uses fourth powers and gives more weight to peak values than rms values.

Griffin (1995) notes that for a sine wave (simple harmonic motion) with a frequency of 1 Hz and a peak-to-peak displacement of 0.1m, the acceleration will be 3.95 ms^{-2} peak-to-peak; 1.97 ms^{-2} peak; and 1.40 ms^{-2} rms (Figure 12.1).

It is important when reporting vibration measurement to include all units and parameter used (e.g. 1.0 ms^{-2} rms). The acceleration due to gravity (g) on earth is 9.81 ms^{-2} and in some, mainly military, studies acceleration if given as a fraction of g (e.g. 0.5g peak, etc.) but this is not now recommended or used.

Vibration can enter the body at many locations and research into whole-body vibration has concentrated on the feet, the ischial tuberocities, when sitting, and the back, when leaning against a backrest or floor. Special consideration is also given to hand-transmitted vibration (see chapters 14 and 15).

For each location, the vibration is considered in three orthogonal directions of translational vibration. These are up and down, side-to-side and fore-and-aft. Rotation around these 'axes' also provides roll, pitch and yaw. The axes of vibration are human-centred so *x*-axis vibration is fore-and-aft when sitting or standing but vertical when lying down, for example. For a sitting and standing subject, *x*-axis vibration

DOI: 10.1201/9781003401964-16

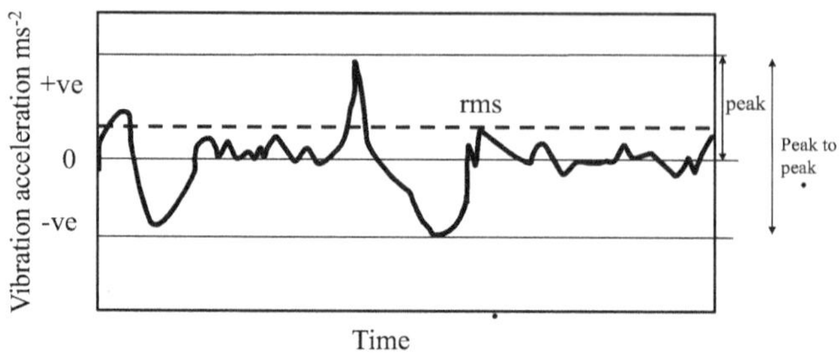

FIGURE 12.1 Vibration time history showing peak; peak-to-peak; and root mean square (rms).

is fore-and aft; *y*-axis vibration is side-to-side and *z*-axis vibration is vertical. Vibration can occur simultaneously in all directions (axes) and locations (feet, seat, back, etc.) and the human response is to the integration of all of these stimuli.

Vibration frequencies in the range from 0 to 100 Hz affect human response to whole-body vibration. If we consider a simple sine wave (simple harmonic motion) the displacement is related to the acceleration divided by the square of the frequency. At low frequencies, the vibration is often visible but for the same acceleration level at higher frequencies, the displacement is small and not visible. So not being able to see the vibration in an environmental ergonomics survey, is not an indicator that it is not present and has no effect.

HUMAN RESPONSE TO VIBRATION

As a general 'picture' for human response to vibration, if we consider vertical (*z*-axis) sinusoidal vibration of a seated or standing person we find that at low frequencies the body moves as one and at frequencies around 0.2 Hz motion sickness is often experienced, exacerbated by voluntary movements of the head. As frequency increases, individual areas of the body begin to move out of phase, notably around 4–8Hz where pelvic and pectoral girdles seem to be 180° out of phase (one goes up when the other goes down).

Above those frequencies, specific areas of the body are affected. Vision begins to be affected by the vibration of the eye at around 10–15 Hz and above that displacement becomes small and vibration is often absorbed near to the point of entry to the body.

For side-to-side (*y*-axis) and fore-and-aft (*x*-axis) vibration, sway will be important at low frequencies around 1–2 Hz, particularly in standing subjects. These responses are not necessarily disturbing as the body is designed to accommodate motion; however, some can be related to effects on health (and safety e.g. balance), comfort and performance. The general description provided above is a rough description useful to provide a basis for understanding when conducting an environmental ergonomics assessment.

WHOLE-BODY VIBRATION FREQUENCY WEIGHTINGS

Subjective methods used to measure the responses of people exposed to vibration, on controlled vibration machines in a laboratory, have produced curves of the relative response of people to vibration frequency for the same level of vibration (investigations have also included combinations of frequencies, complex wave forms and levels as well as axes and input points – see Chapter 2 and Griffin, 1990).

Methods of magnitude estimation have also shown how these curves change with vibration level (Morioka and Griffin, 2006). Curves have been produced for all axes and selected input points on the body (Parsons and Griffin, 1983).

Standardised curves have been derived for use in standards and application and when used as weighting functions (in inverted form with 10Hz given a weighting of 1.0) the weighted rms acceleration (a_{wrms}) value can be derived. Multiplying each vibration input to the body (axis and input point) by the relative sensitivity of the person to that input provides an a_{rms} value for each input and the square root of the sum of the squares of all a_{rms} values provides a final integrated a_{wrms} value for the total vibration exposure.

British Standard, BS 6841 (1987) provides asymptotic (straight line approximation) frequency (f) weightings as follows (see also Chapter 2).

W_b has a value of 1.0 from 5 to 16 Hz; 0.4, from 0.5 to 2 Hz; f/5, from 2 to 5Hz; 16/f from 16 to 80 Hz.

W_c has a value of 1.0 from 0.5 to 8 Hz; 8/f from 8 to 80 Hz.

W_d has a value of 1.0 from 0.5 to 2 Hz; 2/f from 2 to 80Hz

W_e has a value of 1.0 from 0.5 to 1 Hz; 1/f from 1 to 20 Hz.

W_f has a value of 1.0 from 0.125 to 0.25 Hz with a 6dB per octave decrease in sensitivity below 0.125 Hz and a 12 dB per octave decrease in sensitivity above 0.25 Hz.

W_g has a value of $f^{1/2}/2$ from 1 to 4Hz; 1.0 from 4 to 8 Hz; 8/f from 8 to 80Hz

When assessing thermal comfort and ride quality in vehicles W_b is used for *z*-axis vibration at the ischial tuberosities (seat) and *x*-, *y*- and *z*-axis vibration at the feet. W_c is used for assessment of *x*-axis vibration at the back and W_d is used for *x*- and *y*-axes at the seat and y and z-axes at the back. W_e is used in the assessment of rotational vibration in roll (rx), Pitch (ry) and Yaw (rz). The relative multiplying factors for each axis/input are: 1.0 for *x*-, *y*- and *z*-axes and 0.63, 0.4 and 0.2 for rx, ry, and rz, respectively, at the seat; 0.8, 0.5 and 0.4 for *x*-, *y*- and *z*-axes at the back and 0.25, 0.25 and 0.4 for *x*-, *y*- and *z*-axes at the feet.

W_b and W_d are used for the assessment of building vibration; W_f is used in the assessment of motion sickness and W_g is the *z*-axis whole-body vibration weighting curve from ISO 2631 (1985). W_h is a 'realizable weighting' for hand-transmitted vibration. There appears to be no W_a maybe because of confusion with acceleration?

THE VIBRATION DOSE VALUE

The period over which a person receives vibration is termed exposure time, the cumulative amount of vibration is then the dose. Griffin (1995) proposes a vibration Dose Value (VDV) which "can be considered to be the magnitude of a 1s duration of vibration which will be equally severe to the measured vibration". It is given as the integral over time of the fourth power of the weighted acceleration with respect to time all to the fourth root).

The whole ensemble of methods, units, parameters and indices presented above therefore manifest into the VDV as an index that can be related to the human response to any vibration environment and hence used in providing exposure limits for health, comfort and performance. As an aside, Griffin (1995) suggests that a simple approximation to the VDV (eVDV) is $1.4 \times a_{rms} \times t^{1/4}$. The units of the VDV are $((ms^{-2})^4)s)^{1/4} = (m^4s^{-8}s^1)^{1/4} = ms^{-7/4} = ms^{-1.75}$.

British Standard BS 6841 (1987) suggests that high vibration dose values will cause severe discomfort pain and injury but there is no consensus on VDV values that cause injury. A VDV value of 15 $ms^{-1.75}$ will usually cause severe discomfort so values above that can be expected to lead to increased risk of injury. The VDV approach provides a more comprehensive, rational and substantiated approach than the older more speculative method of assessment provided in international standard ISO 2631 (1985), where a time dependency curve is provided for determining whole-body exposure limits.

For an environmental ergonomics survey, measurement and analysis of vibration using frequency weightings and VDV values can be used to provide absolute predictions of the effects of vibration on the health, comfort and performance of those exposed and a relative performance of environments such as those in vehicles. They can also be used to assess environments in terms of contract specifications, regulations, guidelines, standards and limits. For example, a defence contract for a new helicopter may specify maximum vibration limits for exposure of the pilot to avoid interference with vision.

The measurement and analysis of whole-body vibration requires specialist equipment and may not be required in a first-level assessment. Subjective methods will provide a valid and reliable assessment of vibration environments in terms of ride-quality and comfort in vehicles or in buildings, for example, and may provide a first-level indication of influences of performance and health. Guidance is provided in ISO 28802 (2012).

FURTHER READING

Griffin, M. J., 1990, Handbook of human vibration; Academic Press, ISBN 0-12-303040-4

Mansfield, N. J., 2005, Human response to vibration, CRC press, www.crcpress.com, ISBN 0-415-28239-X

STANDARDS

ISO 2631-1, 1985, Evaluation of human exposure to whole-body vibration: part 1 – General requirements. ISO, Geneva.

ISO 2631-2, 1989, Evaluation of human exposure to whole-body vibration: part 2 – continuous and shock induced vibration in buildings (1 to 80 Hz), ISO Geneva.

ISO 2631-3, 1985, Evaluation of human exposure to whole-body vibration: part 3 – Evaluation of exposure to whole-body z-axis vertical vibration in the frequency range 0.1 to 0.63 Hz, ISO Geneva.

HSE, 2005, Whole-body vibration at work, The control of vibration at work regulations 2005.

ISO 10551, 1995, Ergonomics of the thermal environment – Assessment of the influence of the thermal environment using subjective judgement scales, ISO Geneva

ISO 28802, 2012, Ergonomics of the physical environment – Assessment of environments by means of an environmental survey involving physical measurements of the environment and subjective responses of people, ISO Geneva

13 Case Study
Assessment of Vibration Ride Quality in a Petrol and Electric Car

THE ENQUIRY

Gordon was the manager of a car manufacturing company responsible for prototypes and design. The company had produced an electric car that was an updated version of the previous petrol car and wanted to know if they could advertise the new electric car as not only being 'greener' but also with an improved ride-quality. Gordon contacted the Institute of Vibration, Human Factors Department, at the University of Solenti as follows.

Dear Director

I am writing to ask for your help in comparing the ride-quality in our new electric car with that of our petrol version. We have noticed that not only is the electric version much quieter than the petrol version but that it seems to give a smoother ride. Can you please provide us with scientific evidence that this is the case and, if possible, by how much. The objective is to add this to our positive information about the new car to contribute to its advertisement.

Sincerely

Gordon

Head of Design

Michael, the Director of the Institute, passed the enquiry to Jack, the Head of Human Factors and vibration assessment. Jack had much experience in research into human response to vehicle vibration discomfort and its application and had assessed the ride-quality over a wide range of land, sea, air and space vehicles.

THE MEETING

After an international three-centre video conferencing call involving Jack, Gordon and his team, it was decided not to use a test track to provide the data but to use actual roads to measure the ride-quality of the two vehicles. This would then provide realistic data and provide source material for the advertising campaign.

Two cars, one petrol and one electric, with identical outward appearance and interior trim and seats, would be provided by Gordon who would ensure that they were representative of the typical production car and in good order with tyres at correct pressures and so on. Jack emphasised that any results would only apply to the vehicles as presented to the investigation. In a previous investigation, he had assessed two

DOI: 10.1201/9781003401964-17

prototype vans that had arrived with different loads in the rear compartment and that had not been the intention of the client.

For investigations on actual roads, Jack used a standard route (used previously for this sort of testing) with identified sections of road to cover a wide range of vibration inputs to the car. A special request was to include a cobbled street in the assessment, typical of that found in European old towns and elsewhere.

Ride-quality indices would be calculated according to standards and subjective methods would be used. Four males and four females, in good health and from the general public, with no vested interest in the trial, and with a profile (e.g. 20 to 50 years of age) provided by the marketing department, would be used. It was agreed that the ride-quality in the passenger seat would be assessed. This reduced insurance liability and costs and ensured that the assessment involved passengers who would not be distracted by other tasks. Two drivers would be used who had no vested interest in the outcome of the assessment and were both experienced in driving the cars and were familiar with the route and procedures.

Jack noted that although it would be impossible to disguise the nature of the cars, the spirit of the trial would be that there was no difference between the ride-quality of the two cars (null hypothesis) and that it was not expected from the outset that one of the cars had a better ride-quality than the other. This was one reason why it was important that objective measures of vibration ride-quality were taken as well as subjective measures. The two vehicles would be identified as vehicle F (electric) and vehicle G (petrol) where the participants and the team who analysed the data would not be informed of which car was which (they could guess of course, but were not 'allowed' to discuss it).

AGREED AIM AND OBJECTIVES

The following aim and objectives were agreed.

Aim

To compare the ride-quality of two vehicles (one electric car, vehicle F, and one petrol car, vehicle G)

Objectives

Objective 1. To compare the vibration ride-quality indices of vehicle F and vehicle G

Objective 2. To present, comment on, and identify contributions of vibration components, to the vibration ride-quality spectra of vibration inputs to passengers

Objective 3. To use subjective methods to assess ride-quality experiences of passengers and compare the ride-quality of car F and car G

Objective 4. To provide a report to Gordon within one month of the completion of the trial, marked commercial in confidence. It will include an executive summary with a clear conclusion with a clear statement of the relative ride-quality of each vehicle and whether it would be reasonable to state that the electric car has a significantly superior ride-quality to the petrol car

ACHIEVING OBJECTIVES

The two cars were fitted with accelerometers to measure vertical, z-axis, vibration at the feet and on the seat and x-axis, fore-and-aft, vibration at the seat back using appropriate seat mounts and pads to represent vibration inputs to the passengers. Experience had shown that these were the inputs most influential in determining vibration ride-quality in passenger cars.

The passengers were informed that they should focus on how uncomfortable they felt the vibration to be when asked during the ride. They would be driven around a route from 10 am to 11 am, have a break, and then again in the other car from 2 pm to 3 pm.

The two cars were driven in convoy and the survey design balanced passenger experience of a lead car; time of day and gender to reduce any order affects. Two passengers per day completed the trial over a total of four days (Monday to Thursday) with similar traffic and dry weather conditions.

The route took around 1 hour to complete and included stationary but engine switched on; a cobbled street, duel carriageway, motorway section and country road. For each çar, an investigator sat in the back seat with an analysis instrument that recorded integrated a_{rms} (weighted) values over the whole trip. Vibration time histories were recorded at the seat for 30 seconds over an identical part of each road section while maintaining the designated speed. Constant speeds typical of road types were maintained over the road surfaces and were consistent over all trials.

Immediately after completion of the 5-minute engine idling and each of the four types of road section, the passenger was asked to rate how uncomfortable the ride over that particular section of road had been on the scale1. Not Uncomfortable; 2. Slightly Uncomfortable; 3. Uncomfortable; 4. Very Uncomfortable. They also rated on a continuous 10 cm line marked 'Extremely Uncomfortable' at one end and 'Not Uncomfortable' at the other.

The same scales were also used to rate the overall vibration ride immediately after the completion of the route and to answer the question in the afternoon session; in terms of vibration ride-quality which of the two cars did you prefer? These were completed on a single-sheet questionnaire held in a clipboard by the passenger.

The report provided details of the trial and direct responses to ensure that the aim and objectives were met. Additional analysis ensured that order and gender effects were not significant and left some room for answering further questions if required. As engine idling is not part of passenger experience in modern cars, that part of the analysis was not included in the report.

THE REPORT: COMPARISON OF THE RIDE-QUALITY OF A PETROL AND ELECTRIC CAR

EXECUTIVE SUMMARY

The ride-quality in otherwise identical petrol and electrically driven passenger cars was compared, without prejudice. The aim was to determine if it was valid to advertise that one of the advantages of purchasing the electrically driven car is that it provided a better ride-quality than the car that used petrol as fuel. Four male and four female members of the general public, with specified profile, were driven for one hour, around a route on the public highway and vibration was recorded during the trial and while stationary before the Trial. The route included a cobbled street, duel carriageway, motorway and country road over a total distance of 50km. During the trial, subjective ratings of vibration discomfort were recorded and at the end of the trial ratings were provided for the whole route. In addition, at the end of the second trial, direct questions were asked concerning preference of ride-quality between vehicles.

Comparison of vibration indices based upon international standards, and subjective responses of passengers; suggests that it is reasonable to conclude that the electrically driven vehicle provides a better vibration ride-quality experience than the car fuelled by petrol.

COMPARISON OF VIBRATION RIDE QUALITY IN THE TWO VEHICLES

Objective 1. To compare the vibration ride-quality indices of vehicle F (electric) and vehicle G (petrol)

The ride quality index provided in BS 6841 (1987) was used to compare the ride in the two cars. This was achieved from root-mean-square (a_{rms}) values weighted with appropriate frequency weighting curves and multipying factors, for each axis and input of vibration recorded; and taking the root of the sum of squares of the values to provide an overall a_{rms} index. Table 13.1 provides the values for the overall route and for each section of road tested. It can be seen that the electric car had lower a_{rms} values overall and for each section of road. As a reference, a_{rms} values below 0.315 ms^{-2} would be regarded as Not Uncomfortable; 0.315 to 0.63 ms^{-2}, A little uncomfortable; 0.5 to 1.0 ms^{-2}, fairly uncomfortable and 0.8 to 1.6 ms^{-2}, Uncomfortable.

CONTRIBUTION OF FREQUENCY COMPONENTS TO VIBRATION RIDE-QUALITY

Objective 2. To present, comment on, and identify contributions of vibration components to, the vibration ride-quality spectra of vibration inputs to passengers.

The 30-second recorded vibration acceleration time histories for each axis/input and road type were used to determine weighted power spectra. These were presented in

TABLE 13.1
Comparison of vibration ride-quality (a_{wrms} ms^{-2}) in an electric and petrol driven car

Road:	Cobbled street	Duel carriageway	Motorway	Country road
Vehicle F (Electric)				
Floor (z-axis, wb)	0.12	0.22	0.20	0.09
Seat Pan (z-axis, wb)	0.22	0.29	0.25	0.21
Seat back (x-axis, wc)	0.09	0.18	0.10	0.15
Overall Ride quality (awrms)	**0.26**	**0.39**	**0.33**	**0.26**
Vehicle G (Petrol)				
Floor (z-axis, wb)	0.18	0.59	0.25	0.16
Seat Pan (z-axis, wb)	0.22	0.35	0.39	0.25
Seat back (x-axis, wc)	0.12	0.14	0.09	0.10
Overall Ride quality (awrms)	**0.30**	**0.70**	**0.47**	**0.30**

the report. A comparison of the spectra between the two cars showed similarities and differences and the relative contribution of each vibration frequency, axis and input, to the overall ride-quality.

The vibration inputs to the passenger can be interpreted in terms of the operating mechanisms of the car and the road system. The interaction of the road profile, vehicle speed and suspension system provide a low-frequency vibration up to 2 Hz (but not below 0.5 Hz and in the region of motion sickness). Engine revolution rate can also be detected as well as responses of the vehicle seat.

Levels varied with road type in a consistent way between the cars. The most important difference between the cars is that of the engine characteristics where, as expected, the electric car exhibits higher frequency vibrations than the petrol car. This leads to reduced transmission in the electric car through the seat pan (although seat dynamics still apply) when compared to the petrol car where engine and seat characteristics contribute to the ride. Although the fore and aft (x-axis) vibration in the electric car is greater than that in the petrol car, the difference in discomfort is not of sufficient magnitude to cancel out the improvement in overall quality provided by the electric car.

In summary, it can be concluded that the electric car provides a higher frequency input to the passenger than the petrol car and as a consequence the z-axis vibration input to the passenger at the seat pan is greatly reduced, providing overall reduced vibration discomfort. This is most notable on the motorway and least on the cobbled road as the relative importance of the road characteristics and seat dynamics vary.

PASSENGER VIBRATION RIDE-QUALITY EXPERIENCES

Objective 3. To use subjective methods to assess ride-quality experiences of passenger and compare the ride-quality of car F (electric) and car G (Petrol)

The overall ratings for vibration ride quality were on average 2.3 (electric) and 2.8 (Petrol) on the Uncomfortable scale, rating overall slightly uncomfortable for both, with a range of average ratings from Uncomfortable for the cobbled street to Not Uncomfortable for the motorway. A coefficient of harmonisation showed significant ($p < .05$) agreement between passengers and a Wilcoxen Matched pairs signed ranks test showed that the electric car had a significantly ($p < .05$) better ride than the petrol car overall and for all road conditions apart from the motorway where two of the passengers gave less uncomfortable ratings in the petrol car and six the electric car. A simple binomial calculation showed significant ($p < .05$) preference for the electric car in a straight overall comparison.

REPORT: DOES THE ELECTRIC CAR PROVIDE BETTER RIDE-QUALITY?

Objective 4. To provide a report to Gordon within one month of the completion of the trial, marked commercial in confidence and providing an executive summary with a clear conclusion identifying: a clear statement of the relative ride-quality of each vehicle and whether it would be reasonable to state that the electric car has a significantly superior ride-quality to the petrol car

A report was sent to Gordon indicating that it would be reasonable to advertise the electric car as having an improved ride-quality over the petrol car.

THE RESPONSE

Gordon circulated the report to his team, the advertising department and senior managers. A request came back that although the investigation was sufficient to demonstrate an improved ride-quality in the electric car, the objective index used was based upon British standards. An international standard would be required if the results were to match the worldwide advertising that would be required.

Jack defended the use of BS 6841 (1987) as being based upon the most recent and accepted method for determining ride-quality. However, the International Standard, ISO 2631-1 (1985), although based upon less developed information, is the International Standard that is also being adopted as a standard for the European Community. He therefore agreed that confirmation of the results using ISO 2631-1 (1985) would be important and agreed to conduct further analysis.

ISO 2631-1 (1997) provides similar results in most cases to that of BS 6841 (1987). It uses a frequency weighting curve for vertical (z-axis) vibration labelled w_k and differs from the w_b curve (BS 6841, 1987) in that it is slightly less sensitive to vibration below 3 Hz and slightly more sensitive above 12 Hz. Analysis of the data using the w_k curve suggests that the electric car has a less uncomfortable ride than that of the petrol car. This confirms that international standards would support the conclusion that the electric car has an improved vibration ride-quality over the petrol car.

OUTCOME

The report and its conclusions were accepted. The worldwide advertising campaign was then conducted to promote the electric car as a green, sustainable and easy to charge. It was promoted as an improvement on using a petrol car. No mention was made of noise, ride-quality or of any other qualities related to human response to the environment.

FURTHER READING

Griffin, M. J., 1990, Handbook of human vibration; Academic Press, ISBN 0-12-303040-4
Mansfield, N. J., 2005, Human response to vibration, CRC press, www.crcpress.com, ISBN 0-415-28239-X

STANDARDS

BS 6841, 1987, Measurement and evaluation of human exposure to whole-body mechanical vibration and repeated shock. BSI London.
ISO 2631-1, 1985, Evaluation of human exposure to whole-body vibration: part 1 – General requirements. ISO, Geneva.
ISO 2631-3, 1985, Evaluation of human exposure to whole-body vibration: part 3 – Evaluation of exposure to whole-body z-axis vertical vibration in the frequency range 0.1 to 0.63, Hz, ISO Geneva.
ISO 28802, 2012, Ergonomics of the physical environment – Assessment of environments by means of an environmental survey involving physical measurements of the environment and subjective responses of people, ISO Geneva

14 Case Study
Assessment of Vibration in an Office

THE ENQUIRY

Maria was the manager of an office above an underground railway station in a big city. Some of the workers in the office had felt what they called tremors in the office and were concerned about the vibration sufficiently to cause distraction from their work. The office was not in an earthquake zone and any vibration is probably caused by passing trains, a point suspected and suggested by the workers.

To confirm that this was the case, to ensure that levels were within regulations and to determine the level of disturbance, Maria contacted Martin at the Government Building research establishment as follows. Martin and Maria had worked together as PhD students, at the Institute for Perception, at Yalvard University.

Dear Martin

I hope that you are well. I am hoping that you can help me with a problem. I am the manager of a large open plan office, in a building above a railway line. The workers are concerned about vibration in the office, not because it is uncomfortable, although it does seem to be distracting, but mainly because they want to be assured that the levels are within international and national guidelines. Would it be possible to conduct an environmental ergonomics vibration survey to establish the amount of disturbance this may be causing and determine whether the levels are acceptable in terms of regulations?

Many thanks

Maria.

On receipt of the note, Martin explained to Maria that he would conduct a survey involving measurement of vibration at the positions where workers would detect it and also that he would use subjective measures to determine the level of disturbance. A contract was drawn up with the following aim and objectives.

AGREED AIM AND OBJECTIVES

AIM

To evaluate the vibration in an open plan office in terms of accepted regulations and the disturbance it causes to workers

Objective 1. To measure the vibration in the open plan office

Objective 2. To compare the levels of vibration measured with those considered acceptable by national and international standards

DOI: 10.1201/9781003401964-18

Objective 3. To use subjective methods to identify the level of disturbance caused by the presence of vibration in the office

Objective 4. To provide Maria with a report, within one month of the survey, marked 'commercial in confidence'

ACHIEVING OBJECTIVES

Observations and preliminary discussions established that although noise was not an issue, perceptible vibration was linked to the passing of trains. The five-step approach to successful vibration measurement was used (Mansfield (2004) to measure the vibration in the office that represented exposure of workers. That is strategic planning of approach; collation and calibration of equipment; mounting of accelerometers; measurement of the vibration; and analysis and postprocessing. Trains provided a frequent (one every 20 minutes for 30 seconds), regular, but intermittent vibration exposure. Accelerometers were used to measure the vibration on the floor and seat pan (using appropriate instrument 'mounts') at four points across the office, representative of places where people worked and were exposed to the vibration. Eight accelerometers were used, two at each position. They had been calibrated in the laboratory and an additional check was used by turning the accelerometer to provide an indication of the voltage provided for the acceleration due to gravity ($1\ g = 9.81\ ms^{-2}$).

Subjective measures were taken with a one-page questionnaire using an interview technique where Martin completed the questionnaire using the responses for all 30 workstations (workers) in the office. There were 20 people who identified as female and 10 as male. Maria was not present for the interviews.

Martin explained that it was the person's own personal responses that were required, that the questionnaire was concerned with vibration in the office and that any opinions were welcome. Time and location of each interview were recorded and for 15 of the workers (5 male and 10 female) train vibration was present during the interview. The questions were shown to the worker with any associated scales and the worker observed Martin record their responses including any ratings on the scales.

The questions were in four parts: details of the person (and name or identification if agreed), job and location in the office; How the vibration felt now in terms of perception and disturbance with any other comments; How vibration generally felt in the office, any disturbance and any other comments; and a direct question regarding whether vibration was acceptable or not and whether there were any other factors that influenced their working environment with suggestions for improvement. (see Figure 14.1).

Objective 1. To Measure the Vibration in the Open Plan Office

Each of the eight accelerometer signals was recorded throughout the day on a separate channel. In addition, for each location, an a_{rms} value was written down for the period when a train could be detected.

The data were analysed in two parts: periods when no train was present, providing background vibration levels and periods when a train was present, providing higher

Please let us know about your vibration experience in the office. Name or workplace DoB Gender.

Please indicate how YOU feel vibration NOW.	Please indicate how disturbing YOU find any vibration in the office NOW	Indicate why you gave those responses or any other Information about vibration in the office (distracting, affects work etc.)
Clearly	Intolerable	
Slightly	Extremely disturbing	
Just about	Very disturbing	
Not at all	Disturbing	
	Fairly disturbing	
	A little disturbing	
	Not disturbing	

Please indicate how YOU feel vibration GENERALLY in the office	Please indicate how disturbing YOU GENERALLY find any vibration in the office.	Indicate why you gave those responses or any other Information about vibration in the office (distracting, affects work, etc.)
Clearly	Intolerable	
Slightly	Extremely disturbing	
Just about	Very disturbing	
Not at all	Disturbing	
	Fairly disturbing	
	A little disturbing	
	Not disturbing	

Are you generally satisfied with vibration levels in the office? Yes No

FIGURE 14.1 One-page questionnaire about office vibration.

levels for shorter periods of around 30 seconds (and 20 times over the period of measurement), providing a change that may cause distraction and disturbance.

Vibration signals were conditioned to avoid bias (see Mansfield (2004) and analysed for vibration over the frequency range 1 and 80 Hz. It was assumed that vibration of less than 1 Hz (e.g. due to building sway (caused by wind etc)) was not significant (see ISO 6897, 1984).

For the purposes of this study, z-axis (vertical) floor vibration was analysed, leaving data for further analysis if deemed necessary. Background vibration was mainly between 10 and 30 Hz and at 0.025 ms^{-2} rms. Train vibration was almost identical for each epoch and at around 20 Hz Frequency and 0.15 ms^{-2} rms over the mid and highest 15 seconds of the vibration period. The office floor was covered in carpet and although some footfall could be detected it did not seem significant and was not a cause of concern reported by the workers.

Objective 2. To Compare the Levels of Vibration Measured with Those Considered Acceptable by National and International Standards

The absolute perception curve provided in BS 6841 (1987) was used where satisfactory levels above the perception curve for offices are 4× the perception curve value for continuous vibration and 128 times the perception threshold for intermittent vibration.

For background vibration without trains, the magnitude of vibration was 0.025ms^{-2} rms at 10–30 Hz, which is around 2× the vibration threshold value with

75% probability of detection by 50% of fit alert persons (BS 6841, 1987). As this is less than 4×, the value as indicated by accepted national and international standards it can be concluded that background vibration is acceptable. However, it should be noted that any perception of vibration in an office may have affects on people, depending upon a range of factors (e.g. overall satisfaction with work, disposition to be distracted, fear of damage or collapse, etc.).

A value of 0.15 ms^{-2} rms for the intermittent vibration caused by trains is around 10× the threshold value at 20 Hz and hence below the multiplying factor of 128× for offices and intermittent vibration. It can be concluded that the levels of vibration are acceptable according to accepted national and international standards.

Griffin (1990) suggests that the background vibration at the level measured provides 'probable perception'. For the levels provided by train vibration, perception is estimated as strongly perceptible.

Similar results were found when Wb – weighted acceleration – was used to calculate weighted rms values. This is not surprising as it is similar to comparing directly with perception curves at the frequencies considered.

A further analysis of the overall measured vibration for each train calculated the vibration dose value (VDV) value. This can be considered to be the magnitude of a 1 s duration of vibration that would be equally severe to the measured vibration. This was between 0.03 to 0.08 $ms^{-1.75}$ across the office. The total VDV over time is the fourth root of the sum of the fourth powers of the individual VDVs. So for 20 trains this is between 0.063 and 0.169 $ms^{-1.75}$ over the office. This would provide a low probability of adverse comment. At those levels, around 800 trains per day would be required before adverse comment might be expected and standards might consider the vibration unacceptable.

A diagram of vibration levels measured across the office was presented.

Objective 3. To Use Subjective Methods to Identify the Level of Disturbance Caused by the Presence of Vibration in the Office

Although a comparison with accepted national and international standards demonstrated that vibration levels are within accepted limits, subjective methods are more sensitive to context and provide an indication of actual disturbance of vibration to the workers.

It was found that there were 30 (computer) workstations in the office, occupied by 20 females and 10 males of age range 20 to 61 years of age. All wore 'normal' office clothing including shoes and all were seated with feet on the floor, although they could move around the office and in an adjacent kitchen area at will. Observation of the office did not find any particular responses to passing train vibration. Analysis of the responses provided by the questionnaire indicated that all workers were aware of the vibration from passing trains but perceptions ranged from not feeling it to clearly perceptible but no one found it disturbing. Background vibration was found not to be perceptible. Identical questions regarding how the person generally felt at work with regard to vibration provided similar results. It can be concluded therefore that the vibration in the office, although detectable; did not cause disturbance, was acceptable and did not cause interference with work.

General comments on the environment were that passing trains were audible but expected so not disturbing apart from an occasional throbbing; that some workstations suffered screen reflections; that there was sometimes a smell of diesel fumes; and that it was often too hot.

From the subjective survey, it can be concluded that vibration was detected but did not cause dissatisfaction or disturbance to workers. A further, more complete, environmental Ergonomics survey may be useful to provide recommendations for an improved environment.

THE REPORT

Executive Summary

The aim of the environmental ergonomics survey was to evaluate the vibration in an open plan office in terms of accepted regulations and the level of detection and disturbance it caused to workers. Vibration was measured at four sites on the floor and seat at workstations across the open plan office. Subjective responses of all 20 female and 10 male workers were also taken.

It was found that vibration levels were generally above vibration detection thresholds when trains went passed below the building (approximately 20 times per day) but below levels that would cause adverse comment according to accepted national and international standards. Subjective assessment indicated that workers were aware of the trains and some clearly detected the vibration. However it was found to be expected, did not cause disturbance or distraction and was acceptable.

Additional Comments regarding the environment in the office referred to possible dissatisfaction with air quality, noise and the thermal environment. These could be investigated in a more comprehensive environmental ergonomics survey.

A report was sent to Maria within one month of the survey with a supporting video link call to present the findings and take questions.

THE OUTCOME

Maria posted the report with its conclusions on a notice board in the office and confirmed that workers could be reassured that any vibration levels were well within national and international standards. The environmental ergonomics vibration survey had however raised the profile of the office working environment for discussion by workers and provided disappointment and dissatisfaction when Maria revealed that a follow-up study could not be arranged because of budget restraints.

FURTHER READING

Griffin, M. J., 1990; Handbook of human vibration; Academic Press, ISBN 0-12-303040-4

STANDARDS

ISO 2631-1, 1985, Evaluation of human exposure to whole-body vibration: part 1 – General requirements. ISO, Geneva.

ISO 2631-2, 1989, Evaluation of human exposure to whole-body vibration: part 2 – continuous and shock induced vibration in buildings (1 to 80Hz), ISO Geneva.
ISO 2631-3, 1985, Evaluation of human exposure to whole-body vibration: part 3 – Evaluation of exposure to whole-body z-axis vertical vibration in the frequency range 0.1 to 0.63 Hz, ISO Geneva.
BS 6472, 1984, Evaluation of human exposure to vibration in buildings (1Hz to 80 Hz). BSI, London
ISO 28802, 2012, Ergonomics of the physical environment – Assessment of environments by means of an environmental survey involving physical measurements of the environment and subjective responses of people, ISO Geneva

15 Human Response to Holding a Vibrating Object

HAND-TRANSMITTED VIBRATION

When the hand is in contact with a vibrating object, the vibration is transmitted into the hand and often into the wrist, arm and to the rest of the body. This is termed hand-arm vibration or hand-transmitted vibration and it can cause significant discomfort in its early stages but can lead to damage to the blood vessels, nerves, muscles, bones and joints of the hands, wrist and arms.

The consequent chronic pain caused by hand-transmitted vibration and reduction in the ability to carry out tasks can greatly reduce a person's quality of life and ability to work. The damage can be temporary at first but soon becomes permanent and although some moderation may be possible, there is currently no cure.

Although the relationship between hand vibration, symptoms and physiological damage is not fully understood, some guidance on the prediction of the effects of vibration is available and limits have been agreed in national and international standards. If limits are exceeded such that an environment is considered to provide an unacceptable risk of damage to hands, from vibrating tools for example, those responsible for the work as well as tool manufacturers may be liable to pay compensation to workers. As a consequence, an environmental ergonomist may be required to conduct a survey of work involving the measurement of vibration exposure caused by the operation of hand tools, as well as any damage to health that may be caused.

MEASUREMENT OF VIBRATION TRANSMITTED TO THE HANDS

There are a multitude of vibrating hand tools, from industrial chipping hammers, pneumatic drills and sanding and grinding systems to domestic drills and hedge trimmers. We can add to this a multitude of devices from electric toothbrushes, vacuum cleaners and hair dryers to vibrating motorcycle handles.

Tools are held in different ways by different people and by the same person at different times. They may wear gloves and use one or both hands. Measurement of hand-transmitted vibration that is reliable (repeatable for the same conditions) and valid (representative of the actual vibration transmitted) can only be an approximation. The best that can be achieved is to determine standard procedures that will provide a common and accepted frame of reference, as well as clear reporting of how measurements were made.

A starting point is to define vibration axes with respect to the hand (ISO 8727, 1997). For a grasping hand, the third meta-carpel bone on either hand can be taken as

DOI: 10.1201/9781003401964-19

the anatomical origin. The z-axis (Zh) is along the axis of the third metacarpal bone; Xh is anterior through the palm normal to Zh; and Yh is right-to-left normal to Xh and Zh (Figure 15.1).

For a basicentric (tool-based) system when a handle is held horizontally, Xhb is vertical; Yhb is parallel to the long axis of the handle (side-to-side) and Zhb is horizontal (fore-and-aft - normal to Xhb and Yhb). Rotational vibration (Roll, Pitch and Yaw) are defined accordingly but are usually included in any assessment of the translational axes. (Figure 15.2)

Although this sounds complex, the environmental ergonomist requires a (tool-based) system for locating accelerometers to measure vibration in the three axes at the tool-hand interface. A good guide is to place the accelerometers orthogonally with respect to the basicentric system where the (weighted) vibration is maximum.

Weighted root-mean-square (awhrms) values are usually taken as having some validity in predicting the human response to tool vibration, with estimated frequency weighting functions for each axis of vibration (that require improvement). A frequency range of 8 to 1000Hz is often taken and the w_h weighting function (Figure 2.11, Chapter 2), taken for all three axes (See ISO 8041, 1990 and ISO 5349-1, 2001). That is, for sensitivity to vibration acceleration, a weighting of 1.0 from 8 to 16 Hz and a decrease of 6 dB/Octave from 16 to 1000 Hz. The integrated vibration exposure for all axes, (frequencies and levels) was calculated as the square root of the sum of the squares of the awhrms values for each axis. If only the dominant axis is measured then Nelson (1997) suggests that the value is multiplied by 1.4 to provide a practical estimate of multi-axis vibration exposure.

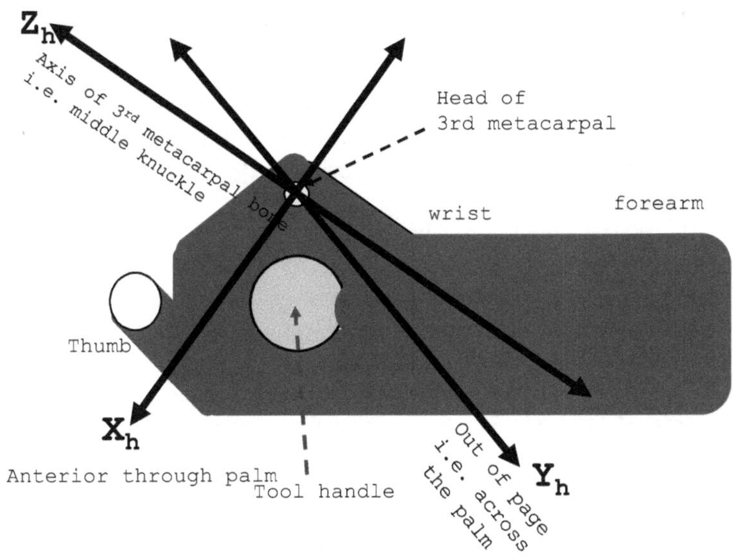

FIGURE 15.1 Schematic representation of the clenched hand showing anatomical hand-transmitted vibration coordinate system. ISO 8727 (1997). Mechanical vibration and shock – human exposure biodynamic coordinate system.

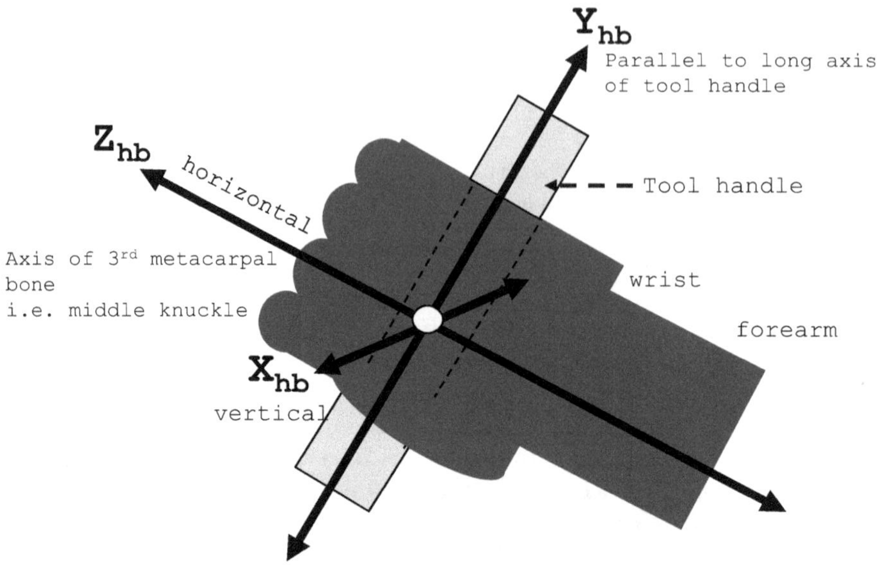

FIGURE 15.2 Schematic representation of the clenched hand showing a Basi-centric coordinate system for the hand holding a horizontal handle.

The effect of vibration duration is not fully understood but can be taken into account by calculating the energy equivalent exposure for an 8-hour day of work. The energy equivalent value for any exposure period is calculated as the overall square root of: the sum of the exposure time (s) × the square of the frequency-weighted acceleration, values throughout the period. That is awh (eq. 8h) for an 8-hour day, etc. (ISO 5349, 2001).

In effect, the time period may be doubled if the vibration magnitude is reduced by a factor of 4. This can lead to excessively high magnitudes to achieve for example a significant 4 h energy equivalent frequency weighted acceleration magnitude if the duration is very short.

Mansfield (2004) provides a review of standards. ISO 5349 (2001) provides general requirements (Part 1) and practical guidance (Part 2) for measurement and evaluation of human exposure to hand-transmitted vibration. Minimum information required includes the subject, operation and nature of the power tools; location and orientation of transducers; the root-mean-square single axis frequency weighted acceleration measured; and total vibration exposure for each operation leading to daily exposure (i.e. 8h energy equivalent daily vibration exposure A(8)). HSE (2005a) provides equations to calculate daily and weekly exposures.

HAND-TRANSMITTED VIBRATION LIMITS

Griffin (1990) notes that limits for exposure to hand-transmitted vibration should be based upon an unambiguous measurement and evaluation procedure, associated with a known probability of an identifiable outcome (adverse effect). He goes on to imply that the nature of the problem is so complex that the best that can be achieved would be reliability (repeatability of measurement and outcome for the same conditions) with some validity (the outcome has some relationship with the actual human response).

Griffin (1990) provides an extensive review of worldwide national and international standards concerned with hand-transmitted vibration and Mansfield (2004) provides a more up-to-date summary as well as descriptions of standards (test methods) for assessing specified hand-held portable power tools (ISO 8662, 1988 to 1996; parts 1 to 14; from chipping and riveting hammers to stone working tools and needle scalers and many in between).

ISO 5349 (2001) tentatively suggests that symptoms of hand-arm vibration syndrome are rare at an A(8) of less than 2.0 ms^{-2} (the European Physical Agents Directive, 2002, suggests an A(8) value of 2.5 ms^{-2} as an action limit). A doubling of rms acceleration, reduces the allowable exposure time by a factor of four. Action limits attempt to provide vibration levels below which exposure is not expected to damage health. At exposures above the action limit, actions must be taken to reduce the exposures to safe levels.

Limits for exposure over a working life are confounded by many lifestyle factors and each individual is unique. Criteria for identifying the damage caused by hand-transmitted vibration as a prescribed disease suitable for worker compensation will depend upon symptoms, years of exposure and other factors that can be argued in court. Attempts have been made to provide a dose-effect relationship over a working life (ISO 5349, 2001) but they must be regarded with caution.

DAMAGE TO HANDS AND ITS DIAGNOSIS

When I used a hedge trimmer to trim my privet hedge I noticed that for a few days afterwards, my hands would exhibit numbness which would eventually recover (including a 'temporary' threshold shift). When I 'retired' as an Emeritus Professor, I took up making garden furniture as a hobby, involving hand drills, jigsaws and sanders. (A covered garden seat for my English rose garden was my forte.) Combined with continued garden maintenance this led to severe burning and numbness of the fingers and palms of the hands, particularly on the dominant (right) hand, that over a few months became permanent (particularly in the morning) leading to tingling of the hands, mild carpel tunnel syndrome, osteoarthritis of the thumb and difficulty in using a computer or cutlery and more. Ceasing to use the tools and some physiotherapy reduced the effect but did not cure it. My hands are tingling as I type.

I mention my own experience as it is often difficult to relate to the severity of a disease simply by describing it. Interestingly, I did not experience vibration white finger which is the usual disease mentioned when considering damage to hands caused by hand-transmitted vibration and raises the point that impacts on the hands without vibrating tools can damage the hands (wrists, etc.) and any effects are often caused by a combination of activities (often called 'workers hands').

Griffin (1990) and Mansfield (2004) provide full descriptions of the effects of hand-transmitted vibration on health. Five areas of damage to the hands are described and it is noted that they can occur in combinations. These are vascular disorders;

bone and joint disorders; peripheral neurological disorders; muscle disorders; and other disorders (whole-body and central nervous system). Hand-transmitted vibration-induced vascular disorders cause vibration-white finger (VWF) and non-vascular disorders are collectively known as hand-arm vibration syndrome (HAVS).

VASCULAR DISORDERS

Primary Raynaud's disease (constitutional white finger) is when fingers go white and numb especially when cold. The cause is not clear but there appears to be a genetic component. Secondary Raynaud's phenomenon is when there is a known cause for finger blanching and numbness.

Vibration-induced White Finger (VWF) is caused by reduced blood supply to the fingers. It is not clear of the exact cause but early symptoms of tingling and numbness are common as well as blanching of the fingers. Symptoms start at the fingertips but with continued vibration exposure, the blanching spreads to the whole finger. The demarcation between blanching and no blanching is clear, so that can be used in the diagnosis of severity. Vaso-constriction in the fingers is a natural part of thermoregulation, to preserve heat at the core of the body, so blanching is most evident when a person is exposed to cold.

There are many factors involved, however, for population exposure, ISO 5349-1 (2001) estimates that for 10% of the population exposed, VWF will occur in $Dy = 31.8[A(8)]^{-1.06}$ years of exposure, where Dy is lifetime exposure in years and A(8) is the daily exposure to hand-transmitted vibration (ms^{-2}). On cessation of vibration exposure, some people recover but most don't.

HAND-ARM VIBRATION SYNDROME (HAVS)

Neurological disorders include tingling in the fingers, leading to numbness and are often the first signs of HAVS. Carpal tunnel syndrome (CTS) can also be due to exposure to hand-transmitted vibration where there is damage to the carpal tunnel system where the median nerve and flexor tendons pass between the bones of the wrist and the transverse carpal ligament.

Bone and joint disorders include damage to the lunate bone of the wrist; bone cysts; and osteoarthritis. There is a debate over whether vibration increases bone density or causes decalcification of bone.

Muscular disorders caused by hand-transmitted vibration include loss of grip, possibly due to incomplete contraction. Damage to tendons, or their synovial sheaths becoming inflamed, and tendonitis (tennis and golfer's elbow) are symptoms as are swelling of the extensor tendon sheaths of the thumb or the flexor tendons of the finger.

Disorders of the whole body related to hand-transmitted vibration include back and abdominal injuries as well as enhanced hearing loss due to an interaction with the effects of noise. Posture and manual handling of heavy vibrating hand tools, may also be contributing factors.

Local vibration to other parts of the body includes vibration to the feet, on vibrating platforms, that have been reported to cause vibration white toe. Thompson et al.

(2010) report a case of vibration white feet (VWF) where a miner who had worked on vehicle-mounted bolting machines for 18 years and experienced foot-transmitted vibration, presented with blanching and pain in his toes.

FURTHER READING

Griffin, M. J., 1990, Handbook of human vibration; Academic Press, ISBN 0-12-303040-4

Mansfield, N. J., 2005, Human response to vibration, CRC Press, www.crcpress.com, ISBN 0-415-28239-X

STANDARDS

HSE, 2005, Hand-arm vibration, The control of vibration at work regulations 2005. Health and Safety Executive, UK, www.hse.gov.uk

ISO 8727, 1997, Mechanical vibration and shock: human exposure, biodynamic coordinate system, ISO Geneva

ISO 8041, 1990, human response to vibration measuring instrumentation, ISO Geneva

ISO 5349-1, 2001, Mechanical vibration. Measurement and evaluation of hand-transmitted vibration – part 1, General guidelines, ISO Geneva

ISO 5349-2, 2001, Mechanical vibration. Measurement and evaluation of hand-transmitted vibration – part 2, practical guidance for measurement at the workplace, ISO Geneva

16 Case Study
Assessment of Vibration at Work in a Dockyard, Involving Vibrating Tools

THE ENQUIRY

Workers in the dockyard at Harbourport used vibrating tools (mainly metal grinders) every day as part of the refit and renovation of ships in dry dock. Noah, from the National Institute of Occupational Health, had been asked to make an overview assessment of health and safety in the dockyard. He had become concerned by the length of time workers were spending using the grinders and hence of their hand-transmitted vibration exposure.

The use of the grinder involved holding it with two hands with the rotating grinding surface pressed against the bare metal. Noah considered that he would certainly not wish to carry out such work over a lifetime and suspected that the workers may be susceptible to damage to their health.

Noah wrote to Hudson, an environmental ergonomist with a PhD in human response to vibration, from the University of Solenti, who was the expert who specialised in human response to vibration at the Navy medical research centre, as follows.

Dear Hudson

I am writing to ask if you will conduct an environmental ergonomics survey of workers using grinding tools at the Harbourport dockyard. I would like to re-assure myself that the work is within acceptable limits and guidelines and if not then I will take appropriate action. Please can you take vibration measurements and compare them with accepted international and national limits. In addition, can you conduct a survey of work to ascertain whether there is damage to the health of metal grinding workers caused by hand-transmitted vibration.

Yours sincerely

Noah

Senior Health and Safety Officer

In a video conferencing communication with Noah, Hudson confirmed that he would be able to carry out the work in two weeks from the present date. Noah noted that it would probably be cold weather and although covered from rain and snow, the workers would probably be exposed to low temperatures.

Three workers would be selected to carry out a typical grinding task and Hudson will measure the vibration, assisted by Charlotte, a trainee apprentice ergonomist and intern from the University of Oxborough who will administer a short questionnaire,

DOI: 10.1201/9781003401964-20

immediately after the completion of a grinding operation, to all 20 workers. Noah noted that each grinding operation lasted about one minute; that all workers identified as male; and that they all worked 8.0 am to 4.30 pm shifts, with usual breaks, from Monday until Friday and not at weekends. He also confirmed that staff turnover and absenteeism was low and that the workers were well paid when compared to other workers in the dockyard.

AGREED AIM AND OBJECTIVES

The following aim and objectives were agreed.

AIM

To determine whether the hand-transmitted vibration exposure of metal grinding workers in the Harbourport dockyard was likely to cause a risk to health

Objective 1: To measure the hand-transmitted vibration of workers when operating metal grinders
Objective 2: To compare the exposure of metal grinders to hand-transmitted vibration with accepted international limits
Objective 3: To determine the internationally accepted allowable daily exposures to hand-transmitted vibration of the workers using metal grinders
Objective 4: To use a self-reporting subjective questionnaire to investigate health affects of all Harbourport workers using metal grinders
Objective 5: To provide a confidential report with recommendations for further action if necessary

A contract was agreed, between institutions, and the work was planned and carried out as follows.

ACHIEVING OBJECTIVES

The five-stage process for vibration assessment proposed by Mansfield (2004) (planning; equipment; mounting of accelerometers; measurement; analysis) was adopted as well as the procedures outlined in ISO 5349 parts 1 (2001) and 2 (2001). Action and exposure limits from the European Directive on physical agents (2002) of 2.5 ms^{-2} and 5.0 ms^{-2} were used to determine allowable exposure times for the work so that they could be compared with actual exposure times. These were integrated into the achievement of objectives as follows.

Note. A metal grinder uses an abrasive wheel to remove unwanted excess material from metal surfaces. In the Harbourport dockyard it is mainly used to smooth surfaces on the ships hull, particularly after welding.

Objective 1: To Measure the Hand-Transmitted Vibration of Workers When Operating Metal Grinders

Three metal grinder workers were selected randomly from a sub-set of 10 available workers. Three orthogonally orientated accelerometers were attached on the tool with adhesive, between each (gloved) hand and the tool at locations judged most likely to cause maximum hand-transmitted vibration. Photographic images of the measuring system were taken. Calibration of accelerometers and amplifier gains etc were carried out to ensure optimum measurement using two vibration meters that provided measurement, conditioning and analysis of the vibration. (Mansfield, 2004).

Each worker performed a typical metal grinding operation on one of three separate areas of a ship's hull (one per worker), and the vibration was measured over a period of 1 minute.

The following results for each tool measured were frequency-weighted (w_h) values of 18.2, 7.6 and 7.2 ms^{-2} rms for x, y and z axis vibration respectively for the right hand and frequency weighted (w_h) values of ; 14.2, 6.9 and 6.3 ms^{-2} rms for x-, y- and z-axis vibration, respectively, for the left hand. That is an overall rss value of 21.0 ms^{-2} rms and 17.0, ms^{-2} for the right and left hands respectively for worker A. For workers B and C, overall rss values were 21.3 and 21.2 ms^{-2} respectively for the right hand and 17.2 and 16.5 ms^{-2} respectively for the left hand. For each worker, the maximum vibration transmitted was via the x-axis of the right hand. The mean rss values over all three operators were therefore 21.1 and 16,9 weighted ms^{-2} for right and left hand respectively.

Objective 2: To Compare the Exposure of Metal Grinders to Hand-Transmitted Vibration with Accepted International Limits

It was considered prudent to take the maximum exposure value of 21.3 ms^{-2} (worker B, right hand) for the calculation of allowable exposure times. Using the European Physical agents directive (2002), 2.5 ms^{-2} is the action limit and 5 ms^{-2} is the exposure limit providing 7 minutes before the action limit is reached and 26 minutes before the exposure limit is reached. (a nomogram is presented in HSE, 2005a for a simple method of 'calculation). The formula for calculating daily (8-hour) exposure time for an action limit for a_{hv} of 21.3 ms^{-2} is $T_{EAV} = (2.5^2 \times 8)/21.3^2 = 50/453.69$ hrs $= 0.1102 \times 60 = 6.6$ rounded to 7 minutes. For the exposure limit replace 2.5 for 5.0 in the formula to get 26 minutes.

Objective 3: To Determine the Internationally Accepted Allowable Daily Exposures to Hand-transmitted Vibration of the Metal Grinders

Based upon the most severe measurement of a root of sum of squares value of 21.3 ms^{-2} (worker B, right hand), an action limit of 2.5 ms^{-2}, and an exposure time of the operation of 1 minute; the time to reach the action limit is 7 minutes hence 7 operations. That is when action must be taken to prevent vibration injury. For exposure limits where no further exposure is allowed, the exposure time is 26 Minutes or 26 operations.

Objective 4: To Use a Self-reporting Subjective Questionnaire to Investigate Health Effects of All Workers Using Metal Grinders

Charlotte produced a two-page questionnaire to investigate the possible effects of hand-transmitted vibration on the health of all 20 metal grinders, immediately after they had completed a grinding operation. As a control group, and on the same day as the survey, she gave a parallel questionnaire to 20 male office workers at the Institute of Naval Research, whose work did not involve the operation of hand tools.

Page 1 of the questionnaire involved details of the worker and work they conducted. Page 2 contained a series of rating scales and opinion boxes supported by outline sketches of left and right hands, palm up, showing palm and all fingers and thumb. The rating scales were 100mm lines labelled at each end from no effect to maximum effect, for sensation; discomfort and pain. Opinion boxes provided the opportunity for further comment, including any inconvenience caused to activities at any time.

The workers were informed that the questionnaire was to ensure that the machine grinding work was not causing problems to their long-term health by comparing their work with accepted international standard health and safety limits. They were also asked to report any affects that using the grinders may be having, especially to their hands, immediately after the grinding operation, at work and also when not at work.

The results showed that, apart from white fingers, reported symptoms did not relate to how long (years) the operators had been involved in work using the tools. The number of grinding operations per day varied from 10 to 30 but on some days no grinding work was required.

All 20 operators felt a tingling of the hands immediately after using the tools, particularly the dominant hand. Fifteen of the operators reported a burning of the larger fingers in the morning after sleep and 12 of those said that any symptoms lasted for only an hour or less. Three of the operators had persistent numbness of the hands and they were also those who had white fingers with 5, 8 and 12 years of work using grinders. Two of the operators reported persistent numbness of large fingers and wrist as well as seizure of a thumb, all indicative of carpel tunnel syndrome and pressure on the median nerve.

By contrast, the 20 workers in the control group at the research institute did not report white fingers and 17 did not report numbness or tingling of the fingers either at work or at home. Three of the workers did report numbness and tingling in the morning, symptoms of carpel tunnel syndrome. The results from the control group provide an indication that any symptoms reported by the metal grinder workers were caused by using metal grinders and that vibration may have been a main factor. A control group where similar tools with weights and postures but not vibration, were employed, was not possible but such factors and their interactions with vibration, should not be dismissed.

Overall it can be concluded that there is evidence that work with metal grinders at Harbourport dockyard causes damage to health and that further action is required.

Objective 5: To Provide a Confidential Report with Recommendations for Further Action if Necessary

A report was sent to Noah within one month of the study.

THE REPORT

Executive Summary

The aim of the environmental ergonomics vibration survey was to determine whether the hand-transmitted vibration exposures of metal grinding workers in the Harbourport dockyard were a risk to health. Three metal grinder workers carried out a metal grinding task on the morning of the 8 February, that lasted for one minute. The temperature was 5 °C at the workplace with an air velocity across the (gloved) hands of around 0.5 ms^{-1}.

The vibration assessment was carried out using ISO 5349 parts 1 (2001) and 2 (2001). Orthogonally placed accelerometers measured the vibration on each hand and the most severe vibration was to the right hand of each worker. The integrated total exposure for the most extreme conditions was 21.3 ms^{-2}. Interpreted over an estimated 30 grinding operations per day (i.e. 30 minutes of exposure) it can be concluded that the exposure to hand-transmitted vibration was above accepted limits and that action was required.

Using an action limit of 2.5 ms^{-2} and an exposure limit of 5 ms^{-2}; an 8-hour equivalent exposure time would reach the action limit at around 7 minutes (7 operations)and the exposure limit at around 26 minutes (26 operations). So 30 minutes (30 operations) of exposure will exceed both action and exposure limits and action is required to redesign the work, in particular, reduce exposure times.

It should be noted that a weighted (w_h) acceleration of 2.5 ms^{-2} for 8 hours is equivalent to 5 ms^{-2} for 2 hours; 10 ms^{-2} for 30 minutes; 20 ms^{-2} for 7.5 minutes etc. An exposure limit of 5 ms^{-2} for 8 hours is equivalent to 10 ms^{-2} for 2 hours; 20 ms^{-2} for 30 minutes and so on.

A self-reporting subjective questionnaire to investigate health affects for all workers using metal grinders, revealed symptoms that did not occur in a 'control' group (office workers) who did not use metal grinders (note that they were also in different working conditions, etc.). Tingling and numbness, of the hands of the metal grinders, were prevalent to varying degrees and three longer-term workers exhibited vibration white finger.

It is recommended that medical checks are made on all workers and a system of health monitoring be put in place. Workers with clear health conditions may have to be re-allocated to other work and all workers should be exposed to no more than 28 exposures per day and less if practicable.

THE OUTCOME

Noah received the report and, in his own report on the overall health and safety at the Harbourport dockyard, suggested that a more thorough survey of the health effects of hand-transmitted vibration shall be undertaken and measures and actions introduced shall be reported back to him within one month.

FURTHER READING

Griffin, M. J., 1990, Handbook of human vibration; Academic Press, ISBN 0-12-303040-4
Mansfield, N. J., 2004, Human response to vibration, CRC Press, www.crcpress.com, ISBN 0-415-28239-X

STANDARDS

European Commission, 2002, Directive 44/EC 2002, Minimum health and safety requirements regarding the exposure of workers to risks arising from physical agents (vibration), Official Journal of the European Communities, L177.
HSE, 2005, Hand-arm vibration, The control of vibration at work regulations 2005. Health and Safety Executive, UK, www.hse.gov.uk
ISO 5349-1, 2001, Mechanical vibration. Measurement and evaluation of hand-transmitted vibration – part 1, General guidelines, ISO Geneva
ISO 5349-2, 2001, Mechanical vibration. Measurement and evaluation of hand-transmitted vibration – part 2, practical guidance for measurement at the workplace, ISO Geneva
ISO 8041, 1990, human response to vibration measuring instrumentation, ISO Geneva
ISO 8727, 1997, Mechanical vibration and shock: human exposure, biodynamic coordinate system, ISO Geneva

Part V

Light and the Visual Environment

Part 5 is in two chapters. Chapter 17 considers radiation; light; photometric units and how to measure and assess visual and lighting environments. Chapter 18 presents a case study of the assessment of the visual and lighting environment in an office.

Vision is a dominant sensory input for people and the visual system has always been of much interest to scientists, artists, engineers and others. The role of the brain in visual perception has provided visual illusions that have long been of interest and amusement. The eye is the specialist organ in the body for receiving and processing light. We have two symmetrically placed eyes on the head that provide overlapping images and facilitate depth vision.

Early theories of vision suggested that light was emitted from the eye and detected the surrounding environment (actually bats do this with sonar so a plausible theory). The Arabic scientist Alhazen in 1039 first suggested that light into the eye causes vision. Rational science from the 18th century has provided systematic investigation and Kepler was the first to compare the human eye to the operation of an early camera (although of course he did not know what a camera was).

Ergonomics and lighting were stimulated by H C Weston and his work on light, sight and work and the perception of light by the system of eye and brain reported by R L Gregory. Hopkinson and Collins provided an integration of vision and lighting in their book 'The Ergonomics of Lighting' followed up by editions of 'Human Factors in Lighting' by P R Boyce, that forms the basis and more, for modern Environmental Ergonomics Investigation of the visual and lighting environment.

Many international and national bodies concerned with the visual environment are dominated by lighting levels and lighting engineering. For this reason artificial lighting is often given a dominant role at the expense of the overall development of the visual environment. It is important that Ergonomics Standards for the visual

DOI: 10.1201/9781003401964-21

environment are not overly influenced by a focus on levels of light and the selection of lighting systems, particularly those that use artificial lighting and energy when natural light may be available and people would prefer that. There is still some way to go before people are placed at the centre of the design of the visual environment and Environmental Ergonomics is a move in the right direction.

Chapter 17 presents the nature of light and the human response to light. Natural light from the sun is described in terms of electro-magnetic radiation and is received by the eye that adjusts its configuration to accommodate the level of light. (Figure P5). Light is radiation detected by the eye and ranges from 380 to 780 nm in wavelength. Photopic vision occurs at higher light levels and allows people to see colour, depending upon wavelength. Scotopic vision does not allow colour but provides a visual scene at low levels of light. Mesopic vision is in between and is less well understood.

People have different sensitivities to different wavelengths of light and idealised sensitivity curves show greatest sensitivity to photopic vision at 555 nm and to scotopic vision at 505 nm. Human response to light is not directly related to radiation level and the idealised sensitivity curves are integrated into photometric units. Luminous flux is light emitted by a source and is measured in lumens (lm). The candela (cd) is light emitted in a given direction and is one lumen per steradian. The illuminance is the light falling on an area and is measured in lux (lx), where 1 lx = 1 lm m^{-2}.

Measuring horizontal illuminance at the workplace is usually the essential starting point in an Environmental Ergonomics assessment of the visual and lighting environment. Additional considerations are glare; flicker; and reflections as well as aesthetics and whether a visual environment complements its function; whether it is

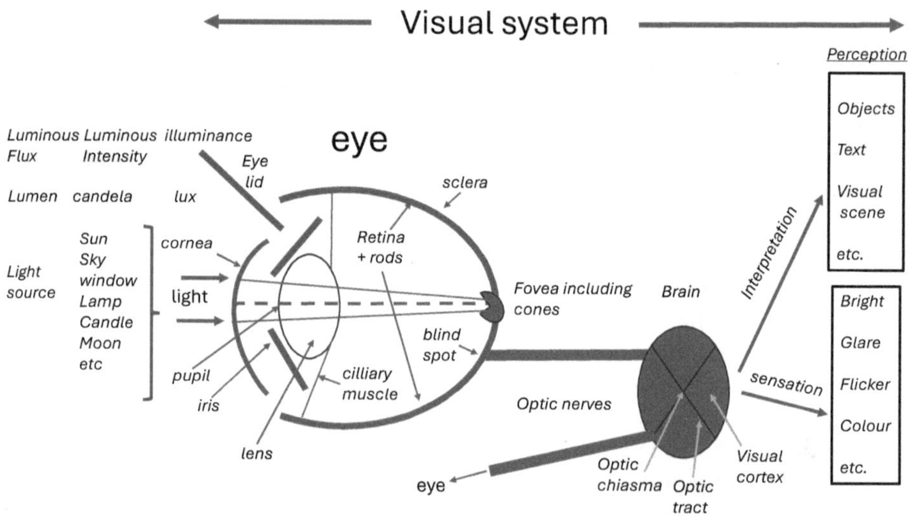

FIGURE P5 Schematic representation of the human visual system. Luminous flux from a source. in a given direction is luminous intensity. Light arriving at a surface is illuminance. Light enters the eye, is conditioned, a coded signal is sent to the brain and it is interpreted as vision.

interesting, stimulating, inspiring and so on. Subjective scales and observation methods are considered along with the RDC (Regulations; Distraction; Capacity) method of assessing human performance, and guidelines for conducting an Environmental Ergonomics survey of the visual and lighting environment.

Chapter 18 presents an Environmental Ergonomics lighting survey of an open plan office. Arthur responded to complaints about the visual environment, among his 30 female office workers, by employing Pedro, a Chartered Ergonomist who specialised in vision and lighting, to conduct a survey. An initial survey involving a light meter and observation checklist concluded that improvements were required. A 5 × 5 grid across the office on an illuminance map, showed light levels from 700 lux near windows to 170 lux towards the far wall. The general impression was functional but in need of cleaning and redecoration.

The room was redecorated over one weekend and a full survey involving subjective, observation and lighting measurements was made. Colour rendering and consistency of luminaires was provided. Levels of illuminance had been raised to a required minimum of 400 lux. 20 of 28 available workers found the visual environment satisfactory (10 of 30 before the redecoration) and up-lighters were recommended instead of spotlights to avoid glare from the back wall.

17 Human Response to Light

RADIATION

The sun emits energy by the fusion of hydrogen to form helium (with 'spare' mass turned to energy), which occurs at a solar-core temperature of around 14 million degrees kelvin. The spectrum of radiant energy is characterised by the nuclear reaction but it is spread by its diffusion through gases and plasma to provide a broad spectrum at the sun's surface temperature of around 5,770 degrees Kelvin.

The energy is emitted in all directions in the form of electro-magnetic radiation that flows through the vacuum of space, and some of it is intercepted by the earth. Light is transferred as photons of energy and it takes around 170,000 years for a photon to travel from the centre of the sun to its release into space and a further 9 minutes to reach earth at a speed of $c = 3 \times 10^8$ ms^{-1} (actually it is a universal constant of 299,792,458 ms^{-1}).

The amount of energy that arrives from the sun at the edge of the atmosphere is around 1373 Wm^{-2} (the solar 'constant'). This varies with the effects of any changes in the sun, intervening cosmic clouds between the earth and the sun and the distance of the earth from the sun which varies. When reduced by scattering and absorption in the atmosphere, the energy arrives, in the form of 'short' wavelength radiation, on the earth's surface in a clear sky at a maximum of around 1000 Wm^{-2}.

The spectrum of solar radiation is broad at the surface of the atmosphere (white in appearance) and is modified due to scattering and absorption by the atmosphere. Some wavelengths are absorbed (providing a yellow appearance in a clear sky and red for example when the sun is lower in the sky and the radiation traverses a greater amount of atmosphere) and scattering depends upon the inverse of the fourth power of the wavelength of the radiation. So short wavelengths perceived as blue are scattered around 9x more than longer wavelengths perceived as red. This provides the blue colour of the sky.

All objects above a temperature of absolute zero, emit and absorb radiation. The higher the temperature, the shorter the wavelength of the radiation. When solar radiation arrives upon objects on the earth therefore, it is not only reflected, but absorbed, increasing the heat content of the object but to much lower temperatures than that of the sun and emitting longer wavelength radiation. The earth, therefore, including all living things, has an environment receiving direct and diffuse solar radiation as well as 'generated' longer wavelength radiation that continually interacts with surfaces and objects.

The electromagnetic spectrum is shown in Figure 17.1. It can be seen that wavelengths vary from cosmic rays of less than 10^{-16} m to radio waves of greater than 10^4 m. The solar spectrum contains wavelengths of 0–300 nm (1.2%); 300–400 nm

DOI: 10.1201/9781003401964-22

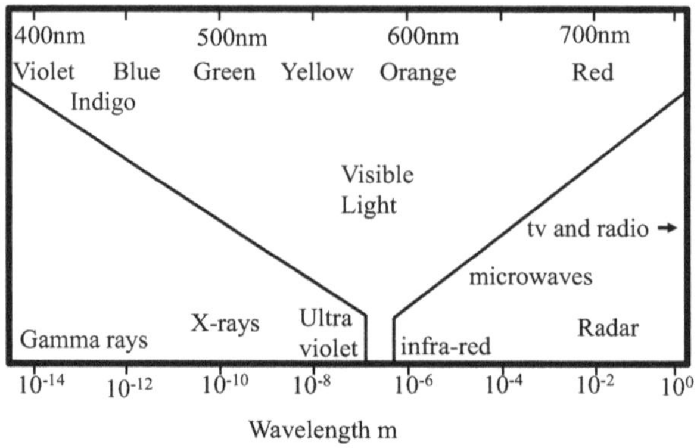

FIGURE 17.1 Electromagnetic spectrum showing the wavelength and spectrum of visible light.

(7.8%; ultra-violet); 400–700 nm (39.8%; visible); 700–1500 nm (38.8%; near infra-red); and >1500 nm (12.4%). Radiation can exist in all forms and life on earth depends upon it and uses it to survive and thrive in the environment.

It is important to understand that light is that part of the electromagnetic spectrum that people have evolved to detect and use to define their world. It is detected and interpreted in terms of its intensity and wavelength, that is its brightness and colour. It should be noted that people also detect other forms of radiation. For example infra-red radiation as heat, so we should perhaps define light in terms of the electromagnetic radiation detected by the human eye.

LIGHT

The human visual system detects electromagnetic radiation over the range 380 nm to 780 nm and this is defined as light. It is important in environmental ergonomics to recognise that human response is not monotonically related to the physical properties of the environment. It is modified by the properties of the 'organic transducer' (the eye), and the perception of the environment (interpreted by the brain). For light this is related to the sensitivity of the photoreceptors of the eye and the signals it sends to the brain, mainly via the optic nerve.

The environmental ergonomist should be aware that this sensitivity is incorporated into the units of light and instruments for measuring light. The eye 'sees' light in terms of brightness and colour (that vary in time and space). The relative sensitivity of the eye to wavelength can be determined subjectively and depends upon the level of light.

An important consideration for the environmental ergonomist, is to understand that vision can be accommodated by the eye over a wide range of intensities. For any given average intensity of light in a visual field, the eye will adapt by varying the amount of light that enters the eye through the size of the aperture (iris) as well as neural and photochemical adaptations.

This provides the eye with a continuously responsive operating range. If a light source falls above the range of adaptation, the light will not be focussed and it will cause discomfort glare as the eye struggles to adapt to the visual field. If light falls below the operating range it may not be seen.

If the eye moves to a much brighter environment, it will adapt within around ten minutes to an appropriate iris size and range (light adaptation). If it becomes darker, adaptation will occur but over a longer period (up to 60 minutes to maximum sensitivity).

Relatively bright light enters the eye, is focussed by the cornea, and is fine-tuned by the lens, onto the fovea, at the back of the eye where there is a concentration of individual sensors that respond to different radiation wavelengths (cones). This is called photopic vision and provides colour as well as brightness.

The relative sensitivity of photopic vision is determined by comparing level and wavelength of light to the equivalent brightness of a standard light at the wavelength of maximum sensitivity (555 nm). A weighting curve is determined by allocating a weighting of 1.0 to the standard light and values less than 1.0 given to wavelengths according to their relative sensitivity. This allows weighted values to be used as units of light and instruments to be 'colour corrected'.

Cones do not respond to low levels of light. Sensors called rods, distributed across the light-sensitive membrane of the eye (retina), integrate their signals across sensors to provide scotopic vision, providing 'black and white (grey) but not colour vision. Using a standard scotopic light at a wavelength of maximum sensitivity of 507 nm, provides a scotopic sensitivity weighting function. The distribution of rods around the retina allows peripheral vision and their density provides the detection of movement. (as an aside, this is what led Griffin (circa 1968) to propose that the reason for blurred vision due to whole-vibration, was the overlapping of an image across sensors on the retina due to a vibrating eye or a vibrating image).

In any environmental ergonomics lighting survey, people involved will vary not only in their sensitivity to the wavelength of the eye but also in their visual performance often involving defects of the eye. Many will have visual defects, often occurring with age, and methods of correction (spectacles; contact lens, etc.). The international organisation for lighting is the CIE (Commission Internationale de l'Eclaraige) and they have produced weighting curves for photopic and scotopic standard observers. (figures 2.4 and 2.5, chapter 2)

Living organisms are influenced by the daily cycle most notably that of night and day. There is an internal 'clock' that provides a cycle and is influenced by external stimuli, particularly light, that stimulates or reduces the production of the messenger hormone melatonin. Light entering the eye is an effective way of influencing the circadian response of the body. The system can be trained using light to adjust the circadian rhythm. The light that influences the circadian response is not part of the visual system but it is transmitted to the pineal gland that produces melatonin.

PHOTOMETRIC UNITS

The radiant flux is the electromagnetic radiation emitted by a source and the flow of energy is measured in watts (W). Luminous flux is the radiant flux weighted over wavelength by the sensitivity of the human visual system (standard observer). It is measured in lumens (lm) by multiplying the radiant flux by 683 lumens/watt for photopic vision and 1,699 lumens/watt for scotopic vision. (based on 1 W = 683 lm for both photopic and scotopic conditions for a radiant flux at 555 nm). In measurements, it is important to state whether photopic or scotopic observers are used and for scotopic conditions to refer to scotopic luminous flux.

Luminous flux is the light given out in all directions. Luminous intensity is the luminous flux emitted per unit solid angle in a specified direction, where a luminous intensity of 1 candela (cd) = 1lumen/steradian and can be used to quantify the distribution of light from a source. The candela is defined in terms of the emission of a black body at the freezing point of platinum (2040 K).

The illuminance is the luminous flux falling on an area, where 1 lm m^{-2} = 1 lux. The luminance is the luminance intensity emitted per unit projected area of a source in a given direction with units cdm^{-2}. Photometric units are described in BS EN 12665 (2011) and CIE (1983).

For the environmental ergonomist, the illuminance (lux), the amount of light arriving on workplaces, is usually the primary measurement made when assessing an environment. Luminance is related to brightness and that along with glare is often assessed subjectively in the first instance, along with the effects of light reflecting off surfaces and aspects of the total visual field. To give a feel for values, daylight in a clear sky can exceed 100,000 lux; good office lighting is around 500 lux and street lighting around 10 lux. (See BS EN 12464-1, 2011; BS EN 12464-2, 2007; CIBSE, 1992; CIBSE, 1994; and Howarth, 2015.)

Limitations of the above CIE system are discussed by Boyce (2014) including spectral sensitivity between photopic and scotopic levels (mesopic) as well as wider fields of view than used to develop weighting curves (standard observers). Individual differences, and especially the effects of age also provide limitations. The CIE system will provide reliability and validity but further consideration may be necessary for specific populations and contexts. (ISO 28803, 2012; ISO 24505, 2016; CIE, 1997).

COLOUR

Light is not inherently coloured, but the interpretation of nervous signals, related to the wavelength and combinations of wavelengths of light, elicit the sensation of colour. Boyce (2014) notes that "colour is a pigment of our imagination" emphasising that it is much influenced by interpretation including past experience.

The reason why an object appears coloured is because its constituents elicit (reflect or transmit) a particular wavelength or spectrum of the light falling on it and it is 'bright' enough to stimulate the photopic system (cones).

The CIE colorimetry system attempts to determine the colour of light using a subjective colour matching system. Although there are three types of cones whose signals are interpreted as colour, the three CIE colour matching functions are mathematical equations (for short, medium and long wavelength radiation) that are used as weighting functions to provide three values related to the imaginary primary colours, X, Y and Z. Using proportions $x = X/(X + Y + Z)$; $y = Y/(X + Y + Z)$; and $z = Z/(X + Y + Z)$ provides three values that can be used on a colour chart to give the predicted resultant colour (Boyce, 2014).

One can only imagine the international research and discussions needed to get this far in producing a reliable and valid system for such a complex but important sensation. Boyce (2014) provides an excellent review. While a reliable system is of paramount importance to the lighting engineer, for the environmental ergonomist, colour, quality of light, aesthetics and other subjective phenomena are best addressed through subjective methods that are the most effective means of determining the holistic effects of light on people.

Two metrics for application have been derived from the complex CIE colorimetry system. The correlated colour temperature uses the system to provide the colour appearance of a light from a light source and the CIE general colour rendering index is related to the effect a light source has on the appearance of surface colours.

The corrected colour temperature determines the temperature of a black body that would give equivalent chromaticity coordinates as the light source. They are usually considered in terms of white light with values of 2700 K towards a yellow warm colour and 7500 K a bluish cool appearance (Figure 17.2).

It should be remembered that the role of the lighting engineer is to provide optimum lighting to an environment and that much consideration has been given to the specification of lighting systems. The power; corrected colour temperature; colour rendering index; and more are therefore specified on luminaires (light bulbs, etc.). The role of the environmental ergonomist is to consider the effects of that environment on the people who occupy it.

Boyce (2014) provides a comprehensive description of the nature of light sources. These are divided into natural light from the sun and sources using flames (candles, oil and gas lamps, etc.). Artificial lighting divides into incandescent lamps that provide light by heating a tungsten filament to incandescence (light due to being heated) and include the light bulb; the tungsten halogen lamp and more, and discharge

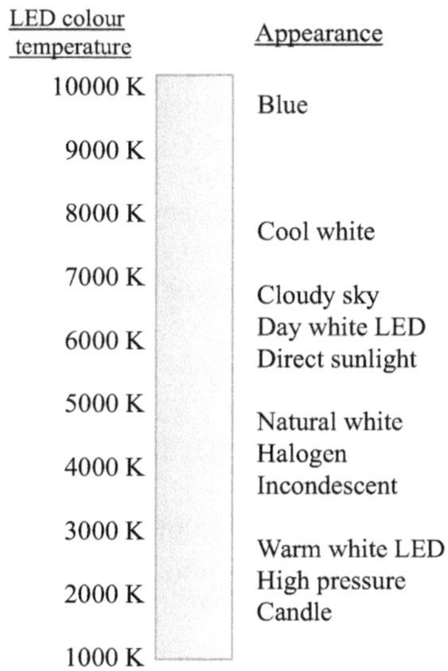

FIGURE 17.2 Colour temperature chart for light emitting diodes (LEDs).

(luminescence) lamps producing light from an electric discharge in a gas and include the fluorescent lamp.

Children are told that there are many sources of natural light: sun; stars; lightning; fireflies; glowworms; bush fires, angler fish and more. When I was young the household coal fire was an important source of light and comfort, supplemented by artificial light and daylight. Artificial light is divided into light emitting diodes; compact fluorescent lamps (CFL); halogen lamps and incandescent bulbs. There are a number of systems for describing the purposes of lighting, including a division into task lighting, accent lighting and ambient or general lighting.

THE VISUAL ENVIRONMENT

As the range of the reception of light to the human eye varies, it adapts so that it can see in environments from a clear sky to a dark night. People are not good at estimating absolute light levels.

The visual system is, however, good at detecting contrast and variation in colour and brightness, over space and time. This determines the field of view for a person. At a different position in the environment, maybe looking in a different direction, the field of view may be different.

The environmental ergonomics lighting survey will use subjective methods as an important way of assessing a visual environment. Objective methods in the first instance will be confined to ensuring that illuminance levels are appropriate for the tasks being performed. For indoor environments, it is usual that horizontal illuminance on the worksurface is measured and compared with accepted standards. The Chartered Institute of Building Services Engineers Code (CIBSE Code, 2022) suggests an illuminance level of 50 lux for areas with little requirement for detail, through 500 lux for general offices up to 2000 lux as the requirement for fine detail increases. CIBSE (1996), provides a lighting guide for visual display units.

SUBJECTIVE ASSESSMENT OF THE VISUAL ENVIRONMENT

There are many facets of a lighting environment that will affect people. As well as psychological aspects including aesthetics, and individual differences and preferences, physical parameters include reflections, flicker, glare, colour rendering, contrast rendering and the direction of light providing modelling affects (an integration of light and shadow to provide form).

The eye can detect oscillating light levels up to the critical fusion frequency (CFF) when it is seen as a continuous light. The CFF is around 10 Hz at low levels of light and up to 50 Hz at higher levels of light depending upon field of view and individual differences. Electric lighting oscillates in level at around 50 Hz and above (differs with different countries) although modern control systems have much increased this frequency. Flicker is almost always undesirable.

Glare occurs when a person experiences bright light above that for which the eye has adapted. There are a number of forms and two of the main outcomes, caused by a wide range of light levels in the luminous field are disability glare and discomfort glare. A practical tip to establish if there is glare, is to block out the possible source of glare (e.g. with a card or hand) and if vision is improved then it is a source of glare.

Colour rendering relates to how people perceive colour in their visual environment. It is not only related to the colour rendering properties of the light sources but also the colours in the visual field, the level of lighting and of course the properties of the visual system with respect to colour of the individual person.

Direction of light can be quantified by vector illuminance (light from a given direction). As the light falls upon an object it will provide modelling effects and shadows that will influence subjective assessments of the quality of the visual environment as well as preferences. An object is seen due to the contrast between the object and its background. In terms of light levels this is the difference between the luminance of the object and the luminance of the background. The contrast may be reduced by veiling reflections. Boyce (2014) notes that the measurement of visual thresholds is influenced by visual system factors; target characteristics and the background against which the target appears. If the contrast is reduced the ability to distinguish fine detail (visual acuity) will decrease.

ISO 10551 (2019) attempts to provide consistency across environmental ergonomics surveys and presents commonly used scales for assessing visual environments. These include:

Perceptual judgements on personal state (after the question "How are you feeling now?" Now, the lighting environment of the room is:)

Extremely dark ______________________________ Extremely light

Evaluative judgements on personal state include 'The visual environment causes me..'

1, no discomfort; 2, slight discomfort; 3, discomfort; 4, much discomfort
Preference scales provide a 'value' judgement. If a subject rates a sensation of 'dark' for example, it does not indicate whether or not he or she *wishes it* to be 'lighter'. The preference rating compares sensation state with how he or she would like to be. No change will indicate acceptability.

ISO 28802 (2012) provides suggestion for the assessment of the visual environment in a first stage environmental assessment. Subjective measures include psychological continua (subjective terms): *visual discomfort*, *preference*, *acceptability* and *satisfaction* as follows:

Visual discomfort scale

"Please rate on the following scale YOUR visual discomfort NOW."

1, No Discomfort; Slight discomfort; 3, Discomfort; 4, Much discomfort

Preference scale

"Please rate on the following scale how YOU would like your visual environment to be NOW."

1. Much darker 2. Darker 3. Slightly darker 4. No change 5. Slightly lighter 6. Lighter 7. Much lighter

Acceptability scale
Acceptable/Not acceptable
Satisfaction scale
Satisfied/Not satisfied
Sources of glare
"Please indicate if you are experiencing any glare NOW."
Observation of the visual environment

The person conducting the environmental survey should also conduct an observation checklist and provide general impressions of the environment. Of particular importance are impressions of the lighting environment, general ambiance, lighting and brightness levels, variations in space and time and sources of discomfort such as glare. Observation of the occupants can indicate adaptive opportunities as well as general feeling for whether the visual environment is optimum. An important consideration is whether the lighting environment complements the purpose of the space.

Howarth (2015) recommends basic equipment as an illuminance meter and a luminance meter (both colour corrected). The average illuminance within a room is measured using a room index (RI) to indicate the number of points to measure. A sketch or blueprint of the room with luminaires marked is required.

The room index is the largest wall length divided by the ceiling height. The minimum number of points to measure (in a grid structure) is (RI:points): <1: 4; 1–2: 9; 2–3: 16; >3: 25. The principle is a guide and in effect it is about statistical sampling to represent the lighting environment across the room. The more points (appropriately distributed) the better. Different shaped rooms can be divided up into squares for example and the RI used to determine the number of points for each square that are then summed to provide the number of points for the whole room.

Howarth (2015) provides a checklist as part of the environmental ergonomics visual survey. He considers problems; measures needed and if guidelines are available. Light levels include light needed and if there is variation across the room and workplaces, and if there are requirements for supplementary lighting for particular areas of the room, times of day and for particular tasks.

Reflectance off walls, ceiling and floor should be noted as well as sources of discomfort and disability glare. Flicker should be noted from fluorescent lights and also computer screens. Colour rendering and requirements are a consideration and also the direction of light and the appearance of objects and shadows. Special features may require highlighting including safety lighting. The particular population who occupy the room and their visual requirements should be considered and, finally, the system for providing maintenance to the visual environment should be noted.

The CIBSE Guide (2022) for lighting is an example of providing guidance for good practice. For example, it suggests that the ratio of minimum illuminance to average illuminance over the task area should not be less than 0.8. (e.g. average of 250 lux, minimum should be ≥ 200 lux). For general offices, the recommended illuminance is 500 lux, more for work with fine detail and less if detail is not essential.

LIGHT AND HEALTH

Ultra-violet and infra-red (UV and IR) light does not affect the visual system but is often present in lighting systems and along with light can have both positive and negative effects on human health. Boyce (2014) provides a review.

The most hazardous light source to which people are consistently exposed is the sun. Radiation can cause damage to the eyes and skin. Visual discomfort can cause eyestrain and damage to health and temporal variations in light can cause migraine and affect those with autism, epilepsy and more. The use of light can help in restoring sleep patterns as well as helping those with Alzheimer's disease and sleep affective disorder (SAD). Light can also be used as a purifier to destroy bacteria, and as a treatment for skin disorders, but it must be used with care.

LIGHT AND COMFORT

Visual comfort and visual discomfort are not part of the same continuum. Guidelines for good lighting are often concerned with the avoidance of visual discomfort, whereas the lighting designer is mainly concerned with providing visual comfort to creating the appropriate mood for the space while promoting functionality and involving for example the creation of stimulating and pleasant environments.

Visual discomfort occurs when performing visual tasks when lighting makes desired information difficult to obtain; there is under- or over-stimulation; and if it causes distraction or perceptual confusion. Visual discomfort can manifest itself as painful eyes and headaches. It can be caused by too much light, too little light; too much variation; disability glare; discomfort glare, veiling reflections, shadows and flicker. (Boyce (2014).

LIGHT AND PERFORMANCE

Visual tasks consist of visual, cognitive and motor components. If the visual component is important then lighting will influence the outcome. Illuminance level is often influential in visual performance (and can improve it up to a point) along with size, luminance contrast and colour difference.

Other generalities about the effects of light on human performance are difficult to make and depend upon context, the task and the individuals conducting the task. Parsons (2019) and ISO TR 23454 – 1 (2022) suggest the RDC (Regulations; Distraction; Capacity) system for predicting the reduction in performance from optimum environmental conditions.

For lighting environments it is unlikely that regulations will cause a limit to work time so time on work is 100%; Distraction may occur due to glare or flicker for example so time on task may be reduced and a reduction in capacity may occur due to non-optimum task lighting. $R \times D \times C$ will then provide an indication in loss of performance at work from performance in optimum conditions. (e.g. if $R = 100\%$ time at work; $D = 90\%$ time on task not distracted; and $C = 90\%$ of the capacity in optimum conditions; then performance is $1 \times .9 \times .9 = 81\%$ of the performance achieved if the environment provided optimum visual conditions).

GUIDELINES FOR A LIGHTING SURVEY

There are many codes and guidelines for lighting from international standards and institutions to national guidelines and more. The Canadian Centre for Occupational Health and Safety (ccohs.ca/oshanswers/ergonomics/lighting) provides useful advice and materials for ergonomics lighting surveys. They note that common lighting problems are insufficient light; glare; improper contrast; poor light distribution; and flicker. Shadows can also be a problem where objects or the worker's themselves can block light.

Survey measurements suggested are average illuminance throughout the workplace; look for shadows; and ask workers if they suffer eye strain, squint or headaches while they are sitting in their workplaces. The person conducting the survey should look for reflected (e.g screens) and direct (e.g. luminaires, the sun, and so on) glare. As noted above, a method of detecting glare is to block the possible source of glare and see if distant objects are easier to see. Suggestions for detecting improper contrast and poorly distributed light are also provided along with practical methods to solve lighting problems.

A case study of an environmental ergonomics survey of the visual environment in an office is provided in the next chapter.

FURTHER READING

Boyce P. R., 2014, Human factors in lighting (3rd Ed), CRC Press, Taylor & Francis Group, London and New York

Howarth, P. A., 2015, Assessment of the visual environment, In Wilson, J. R. and Sharples, S., Evaluation of human work (4th Ed), CRC Press, Taylor & Francis Group, Boca Ratan, New York, Oxford, ISBN 978-1-4665-5961-5, Chapter 25, pp 677–704.

STANDARDS

BS EN 12665, 2011, Light and lighting: Basic terms and criteria for specifying lighting requirement, BSI, London, UK.

BS EN 12464-1, 2021, Light and lighting: lighting of workplaces. Indoor Workplaces, BSI, London, UK.

BS EN 12464-2, 2007; Light and lighting: lighting of workplaces. Outdoor Workplaces, BSI, London, UK

CIBSE,1992, Lighting Guide, The outdoor environment, Chartered Institution of Building Services Engineers, London, UK

CIBSE,1994, Code for interior lighting, Chartered Institution of Building Services Engineers, London, UK

CIBSE, 2022, Lighting for offices, CIBSE Lighting Guide 7 (LG07), Chartered Institution of Building Services Engineers, London, UK. ISBN 9781914543463.

CIE, 1983, The basis of physical photometry, CIE publication 18.2 CIE (Commission Internationale de l'Eclaraige), Vienna.

CIE, 1997, LOW vision lighting needs for the partially sighted, CIE TR 123, CIE (Commission Internationale de l'Eclaraige), Vienna.

ISO 28803, 2012, Ergonomics of th physical environment – Application of international standards to people with special requirements, ISO, Geneva.

ISO 24505, 2016, Ergonomics – Accessible design – Method for creating colour combinations taking account of age-rlated changes in human colour vision, ISO, Genva.
ISO TR 23454-1, 2019, Human performance in physical environments. Part 1: A performance framework. ISO, Geneva.
ISO 10551, 2019, Ergonomics of the physical environment-Subjective judgement scales for assessing physical environments, ISO, Geneva
ISO 28802, 2012, Ergonomics of the physical environment – Assessment of environments by means of an environmental survey involving physical measurements of the environment and subjective responses of people, ISO, Geneva.

18 Case study
Environmental Ergonomics Lighting Survey of an Open Plan Office

THE ENQUIRY

Arthur is a manger of an open plan office where 30 female workers use computers, but fine detail is not required. He has conducted the required workplace assessments and can be confident that the computer workstations and tasks meet requirements. Some workers however are complaining about the quality of the lighting in the environment in terms of general dissatisfaction, discomfort and headaches. He contacted Pedro, a Chartered Ergonomist and also a fellow of the Chartered Institute of Building Services Engineers (CIBSE) to ask for advice.

Dear Pedro

I am writing to ask for your advice and assistance in providing an environmental ergonomics lighting survey of an open plan office for which I have responsibility. The office is roughly rectangular with a 20 m long wall and a total of approximately 60 m^2. There is an even distribution of computer workstations, some near an expanse of windows facing north on the third floor of a five-storey building.

All of the workers in the office identify as female and some are complaining that they are dissatisfied with the lighting environment. It does not affect their ability to do their work it seems but it is distracting. Some also complain of headaches. We have a building services department and a lighting engineer maintains the lighting in the building mainly by changing the fluorescent light bulbs on request.

Any advice would be much appreciated especially any that would enable me to assure the workers that lighting conditions are within agreed codes of practice.

Yours sincerely

Arthur

Manager.

THE MEETING

A video conversation was held between Pedro and Arthur who outlined the work of the office. In effect, it provided international support to an Artificial Intelligence (AI) chat line for selling events. No telephones or direct audio or visual communication with customers was involved.

The workers monitored (up to ten at a time) 'conversations' between the AI avatar and the customer and intervened when they perceived that they were required

DOI: 10.1201/9781003401964-23

(all transparent to the customer, that is, without customer awareness). Operators helped with any enquiries and to move towards selling a product (experience – matched to the customer requirement, for example from hot air balloon experiences to racing on ice in fast motor cars to trips to the Himalayas). They carried out an 8-hour shift with normal breaks as convenient to the customer service from 9.0 am to 5.0 pm (other international offices covering different time zones).

Pedro noted that although often termed a lighting survey, his main focus would be on the health and comfort of workers in terms of the effects of light in the visual environment at each workstation and across the office. It would include both good lighting practice as well as the actual perceptions and responses of workers.

Pedro and Arthur agreed to meet at the office on a day when the workers were not present but the office was set up as if they were. It was agreed that the environmental ergonomics lighting survey would be conducted by Pedro in the first instance by measurement and checklist. A follow-up study with the workers present would be conducted using subjective scales, behavioural observations and spot-check measurements.

AGREED AIM AND OBJECTIVES

The following aim and objectives were agreed.

Aim

The aim of the environmental ergonomics lighting survey was to make recommendations to ensure that the lighting environment in the office meets good practice for the health and comfort of workers.

Objectives

Objective 1. To provide an expert assessment of the lighting environment in the office using a checklist method

Objective 2. To measure the illuminance levels in the office and compare them with good practice

Objective 3. To evaluate the lighting in the office using subjective methods after initial recommendations had been implemented

THE INITIAL LIGHTING SURVEY

Pedro and Arthur met on a Saturday morning with heating and lighting normal for the office, but no workers present. Weather conditions were dry with some cloud. The office was open plan with separate desks but not compartmentalised (no cubicles).

An illuminance meter (colour corrected) had been calibrated in the laboratory and was used to measure horizontal illuminance throughout the office. A blueprint of the office showed the position of workplaces, windows, luminaires and heating systems as well as dimensions.

The room index (RI) was calculated with wall length of 20m: ceiling hight of 4 m (20/4 = 5), so RI = 2–3 suggesting a minimum of 16 points needed. 25 were actually

Observation checklist for the visual environment. To be completed by the investigator in the office (may be use Dictaphone if appropriate).

General impression of the visual environment: One sentence and one word descriptors.

What is best about the visual environment in the office ?

What is worst about the visual environment in the office ?

Would you say that the office was bright, dim, suitable for the task, pleasant, acceptable, of high or low quality as an office space

Looking around the room would you say that there are better areas and worse areas for work in the visual environment ?
Where would you prefer to be located?
Can you identify glare, flicker or reflections that may cause problems ?

Make a note of the activities people are carrying out in the office for each workplace.

Make a note of the spectacles worn by each person in the office and note their location

Note personal adaptive opportunities for people to achieve optimum visual/lighting environment

	Opportunity	Opportunity taken
Adjust view		
Move around		
Move out of the office		
Change workplace		
others		

Note technical adaptive opportunities for people to achieve visual comfort

	Opportunity	Opportunity taken
Adjust lighting		
Adjust workplace conditions		
Open or close window blinds		
Adjust amount of natural light		
others		

Does the lighting/visual environment cause discussion or distraction?

Overall conclusion: Is lighting/visual environment optimum ? Yes/No

FIGURE 18.1 Observation checklist for the visual environment in an office to be completed by the investigator (some observations require occupants to be present).

taken, measured at a height of 1.2 m not shielded and at the centre of an even grid system. This was supplemented by measures on the desk at each of the 30 workplaces.

Illuminance measures were taken with lights on from 10.0 to 11.30 am and 2.30 to 4.00 pm on 22 March 2024 (spring). When the illuminance measure was taken Pedro completed a checklist related to the visual environment at that workplace. In addition, a general impression checklist was completed from each end of the room (see Figure 18.1).

RESULTS OF THE INITIAL SURVEY

The illuminance levels at the centre of the 5 × 5 = 25 grids were shown on the blueprint of the room and contour diagrams of the illuminance levels were also presented (Figure 18.2).

The illuminance levels at the centre of the grids ranged from 655 lux near the windows to 200 lux away from the windows providing an average of 390 lux and a minimum-to-average ratio of 0.51 in the morning. For the afternoon measures, levels ranged from 750 to 150 lux across the grid with an average of 450 lux and a minimum-to-average ratio of 0.38. The maximum horizontal illuminance for a workstation in the morning was 653 lux and the minimum was 150 lux as some shielding occurred from filing cabinets. This increased to 750 and 160 lux respectively in the afternoon.

						Mean	SD
	655	650	700	700	750	691	41
	600	550	500	520	530	540	38
	550	400	400	500	450	460	65
	250	210	205	190	200	211	23
	200	150	200	175	230	191	30
Mean:	451	392	401	417	432	419	
SD:	210	214	211	228	227		200

FIGURE 18.2 Horizontal illuminance lighting levels in the office (lx) presented as values to the centre of an evenly spaced 5 × 5 grid. The means and standard deviations (SDs) indicate the distribution of light across the space.

The general impression of the office was functional but dingy with adequate space but not spacious. Workstations looked cared for but the general space was neglected. Windows were not clean, luminaires were dusty and the paintwork was not inspiring. The colour rendering of the luminaires seemed to be set at levels ranging from warm white to cool blue with no consistency. There seemed to be a gradient of light levels from windows to the non-window side of the room.

The checklist at the workplaces suggested that more light was needed away from the windows and that there was unpleasant variation in tone and light across the room, complemented by the faded magnolia walls and off-white ceiling allowing limited reflectance. Floors had blue carpet in reasonable condition and well cleaned. No discomfort or disability glare was found at any of the workplaces as well as no flicker. The windows faced north and there was a perceived increase in light at the windows in the afternoon.

It was recommended that to improve the lighting in the room a redecoration of the room was required as well as cleaning of windows and luminaires and painting of walls and ceiling. It was also recommended that supplementary lighting be provided to the part of the room where there were no windows and that all workstations should have at least 500 lux horizontal illuminance at each workplace. A general guide of 150 lux at the walls and 50 lux at the ceiling should be taken. As a change was required there was little point in conducting a follow-up survey in current conditions.

An interim report was provided to Arthur within a few days of the survey including full details and diagrams. A verbal report and presentation was also provided by video conference where Jim, the lighting engineer, was present as well as Ali, a senior manager responsible for building services. It was agreed that the room would be refurbished over one weekend, to limit loss of work time.

All recommendations were accepted including all lighting to be identical fluorescent tubes set at a colour temperature 4000 K (cool white suitable for office environments) and supplementary lighting to be provided where illuminance levels were low. One month was left to allow odours to settle and then what was in effect a 'post-occupancy' survey was carried out.

THE FOLLOW-UP SURVEY

Sheila the workers representative, was not consulted, but was kept informed of the actions so that she could re-assure the workers that a full assessment of the lighting was being carried out and comprehensive improvements were being made on the advice of an external expert.

The follow-up environmental ergonomics lighting survey was carried out with a single-sheet questionnaire including rating scales, given to all workers. They completed the questionnaire around the same times as the previous measurements had been made and it was emphasised that it was their own individual responses and opinions that were required and not a collective view. Two workers were 'off sick' (for reasons not directly related to vision), so 28 in total. On reflection, it was decided to repeat the initial measurement and checklist procedure for comparison purposes. Hence providing a full lighting and visual environment survey at one time.

The questionnaire was in three parts and was completed by the worker in their normal field of view. Information concerning identification of worker and workplace was taken as well as age, gender identification, length of time working in the office and at that workplace and any health conditions as well as visual correction systems (spectacles; contact lens').

Rating scales were constructed as 10 cm lines labelled as extremes at either end (not present to extremely present, or opposite extremes such as extremely light and extremely dark); scales with Not; slightly; the term; and very; as qualifiers and a final acceptable or not acceptable forced decision. It was made clear that it was not the software image on the computer screen that was being assessed.

The request was – Please indicate how YOU feel about the visual and lighting environment in the office from YOUR perspective. Make a mark that crosses the line in one place, at a position on the following scales. Answer ALL questions.

These are important points as I found that it is almost always the case that some responders tick to the side of the scale or make one mark that crosses the line twice so a judgement has to be made about intent. Also, it is common that some of the responders fail to complete some of the questions.

The questions were supplemented with a text box that allowed further opinions and points to be made. The questionnaire was administered to all workstations at the same times. It was concluded that although this was in itself a visual task, workers would still be able to give valid responses about their lighting environment. The full questionnaire is provided in Figure 18.3.

The refurbishments included cleaned windows and cleaned and colour-adjusted open fluorescent lamps. The floors and ceiling were painted matt brilliant white and supplementary spotlights were placed on the ceiling near the wall away from the window.

RESULTS OF THE FOLLOW-UP SURVEY

Objective 1. To Provide an Expert Assessment of the Lighting Environment in the Office Using a Checklist Method

Please indicate how YOU feel about the visual and lighting environment in the office from YOUR perspective.
Make a mark that crosses the line in one place, at a position on the following scales. Answer ALL questions

Extremely bright	____________	Extremely dark
Extremely satisfactory	____________	Extremely unsatisfactory
Extremely pleasant	____________	Extremely unpleasant
Extremely good colour scheme	____________	Extremely bad colour scheme
Extremely high Quality appearance	____________	Extremely low Quality appearance

Very uncomfortable	Very interesting
Uncomfortable	Interesting
Slightly uncomfortable	Slightly interesting
Not uncomfortable	Not interesting
Much glare	Very inspiring
Glare	Inspiring
Some glare	Slightly inspiring
No glare	Not inspiring

Please provide any further information about your vision and lighting environment including good and bad points, flicker, poor areas of the room, causes of eye stain or any suggestions for improvements.

Are your satisfied with the type of artificial light and lights in the office?	Yes	No
Does your visual environment contain natural light (e.g. from windows)?	Yes	No
Overall are you satisfied with your vision and lighting environment?	Yes	No

FIGURE 18.3 One-page subjective questionnaire for the visual and lighting environment.

Pedro positioned himself at each workstation and assessed the visual and lighting environment. He found that the whole environment had a brighter and fresher ambience. The light nearer the window appeared brighter than inside the room but he found it not to be disturbing. Colour rendering seemed appropriate but there did seem to be some veiling reflections in some computer screens. The workstations in view of the spotlight may cause discomfort glare. Some of the workstations at the windows also suffered from reflections.

It was decided that the findings of the checklist would be integrated with the results of the questionnaire to provide overall recommendations.

Objective 2. To Measure the Illuminance Levels in the Office and Compare Them with Good Practice

The follow-up survey showed increased lighting levels across the room, despite slightly duller weather conditions than those for the initial survey. The illuminance levels at grids near the windows were around 670 lux during the morning and 700 lux in the afternoon, reducing across the room to around 400 lux in both morning and afternoon. The minimum to average value was therefore around 0.6, however, illuminance levels were at about those for good lighting practice across the room (500 lux) so practical steps such as adding transparent blinds to the windows and reducing the height of filing cabinets may be options.

Objective 3. To Evaluate the Lighting in the Office Using the Subjective Responses of Workers, After Initial Recommendations Had Been Implemented

20 of the 28 workers found the lighting environment to be improved. They answered the questions about the general feelings of lighting at work but suggested in comments that the improved lighting invalidates many of the negative comments.

In addition, the opportunity had been taken to re-arrange allocation of workstations for six of the workers on request. (mainly to sit next to a friend or move away from someone!). Some split the responses into before and after the refurbishment. There was general satisfaction after the refurbishment with some specific points about smells, glare, reflections and reduced privacy.

Ten workers reported noise and 12 reported thermal discomfort and draughts. Seven of the workers said that headaches were still present, especially in the evening after work. Three of the workers reported glare in the new system and had tried to use shielding to blot out the spotlights. Reflections from lamps and windows were reported at five of the workstations.

Overall the visual environment seemed to have been significantly improved with 20 of the 28 workers reporting that it was acceptable whereas only 10 had found it acceptable before the refurbishment. The lighting levels were reported as acceptable and the room generally bright. Walls, ceiling and floor were fresh and bright but there was some discomfort glare from the spotlights (that did not cause disability glare). The windows and lamps caused some reflections in computer screens. There were no reported problems with colour, flicker or changes in space and time.

Twenty of the workers had visual support systems. Five with contact lenses and fifteen with spectacles, ten specifically for reading the computer screen. Workers who had complained of insufficient light had been placed near a window. Ages ranged from 20 to 60 years with a work time in the office from 1 week to twenty years but an average of only two years. A full report with diagrams of the office showing the light levels as well as subjective responses at each workstation, was sent to Arthur. The executive summary is provided below.

EXECUTIVE SUMMARY

The aim of the environmental ergonomics lighting survey was to assess the lighting environment in the open plan office and to make recommendations to ensure that the lighting environment meets good practice for health and comfort and was acceptable to workers. An interim report provided details and findings of an initial environmental ergonomics lighting survey of the office without workers present. The measurement of horizontal illuminance throughout the room and an expert checklist assessment, concluded that improvement of the lighting conditions and visual environment of the workers was required. A refurbishment was agreed and carried out.

A post-occupant (follow-up) lighting survey was carried out one month after the refurbishment. The checklist and measurement of illuminance levels were repeated and in addition a questionnaire was administered to all workers. Lighting levels (illuminance) were found to be much improved and in line with good practice, although some fine-tuning is required to improve light distribution. Supplementary lighting with spotlights increased illuminance levels at workstations, with a gradient from around 750 lux to 400 lux from window wall to the opposite wall without windows. Subjective measures indicated that the lighting environment was much improved, although some workstations now had reflections from lamps or windows and glare from spotlights.

It was concluded that the lighting environment was much improved and close to illuminance levels recommended for general offices by CIBSE lighting code (2022). That is 500 lux with a deviation of 0.8 or no less than 400 lux.

It is recommended that some further design features were required. These include semi-transparent blinds at windows that could be adjusted; diffusers on lamps; replacement of spotlights with uplighters and a reorientation of computers to ensure that there were no reflections. The implementation of the new design features should be taken in a coordinated way with workers confirming their efficacy and acceptability.

FURTHER READING

Howarth, P. A., 2015, Assessment of the visual environment, In Wilson, J. R. and Sharples, S., Evaluation of human work (4th Ed), CRC Press, Taylor & Francis Group, Boca Ratan, New York, Oxford, ISBN 978-1-4665-5961-5, Chapter 25, pp. 677–704.

STANDARDS

CIBSE, 2022, Lighting for offices, CIBSE Lighting Guide 7 (LG07), Chartered Institution of Building Services Engineers, London, UK. ISBN 9781914543463.

ISO 28802, 2012, Ergonomics of the physical environment – Assessment of environments by means of an environmental survey involving physical measurements of the environment and subjective responses of people, ISO, Geneva.

Part VI

Environmental Ergonomics and Air Quality

Part 6 presents two chapters on the environmental ergonomics assessment of air quality. Chapter 19 considers the nature of fresh air, air quality in general and perceived air quality. It presents units and methods of assessment as well as national and international limits for ventilation, carbon dioxide and other contaminants. Chapter 20 presents an Environmental Ergonomics case study where the air quality in a college dining hall was assessed and recommendations made.

Breathing air is an obvious vital human function and air quality must always have been a consideration. People lit fires to keep warm and must have existed in smoky dwellings, with various other smells and pollutants, up until the discovery and wide use of the high chimney in the 12th century in Europe.

It has long been understood that poor air quality ('foul air') can affect health and until the discovery of germs in the mid-19th century was regarded as a cause. That is why the Victorians went to the seaside or mountains for health and the German Scientist Max Von Pettenkofer swallowed cholera bacteria (as it happens at great risk and some illness) to disprove the germ theory of the time. His pioneering work on hygiene included recognition of the importance of monitoring carbon dioxide level as an important measure of air quality. His recommendation of 0.1% CO_2 as an acceptable standard for good room air is still accepted today.

Bio-effluents from a person decay in around 15 minutes in rooms if not renewed, but they were the focus of criteria for ventilation in the 1930s with Yaglou from the USA providing ventilation rates related to density of people and hygiene levels. The ventilation rates were found to be effective but the prevalence of tobacco smoke and

DOI: 10.1201/9781003401964-24

other more modern chemicals that did not decay so easily led to sick building syndrome and Fanger (1988) to propose new units for quantifying odours and assessing air quality. These are used in environmental ergonomics assessments of air quality where subjective ratings of 'smelly' and dissatisfaction are employed along with measurement of CO_2 levels.

Chapter 19 defines and describes what is meant by air quality, perceived air quality and how it is measured. Fresh air contains 78% nitrogen; 21% oxygen; 0.03% carbon dioxide and 0.97% argon and other gases. Air Quality is defined as a deviation from fresh air and typical contaminants (pollutants) are discussed. Some contaminants have odour and some not and some can cause damage to health depending upon dose.

The domain of environmental ergonomics is restricted to odours and smells. Possible direct effects on health would be reported to a specialist in the contaminant. The environmental ergonomics survey would include subjective and observational methods. A panel of independent participants may be used to assess environments as instruments are often not sufficiently sensitive. Measurements of carbon dioxide levels are recommended, both as an important output from respiration as well as an indication of the effectiveness of ventilation in reducing pollution in general.

The nose and mouth contain specialist sites for detecting odours and smells. Taste and smell are intricately linked. Saturation of sensors leads to adaptation so the perceived intensity of a smell by a person reduces with time in a space. Air enters the nose and smells are detected in the olfactory epithelium, react with chemoreceptors and signals are sent to the brain for interpretation via the olfactory nerve (Figure P6). The sensitivity of the nose is well beyond that of physical instruments and the interpretation by the brain is complex, allowing the identification of thousands of odours and smells.

The research of Fanger (1988) and Fanger et al. (1988) is described. The olf is a unit of odour intensity and the decipol of pollution it causes (depends upon ventilation rate). Surveys of buildings are presented that show that pollution sources are not dominated by people as previously assumed but that the totality of perceived odour intensity could be related to the number of nominal standard persons (olfs) (Figure P6). Environmental ergonomics methods for the assessment of air quality are presented. A subjective scale of 'smelly' is described and percentage of dissatisfied persons is used to determine the decipol value. Air quality standards are also presented.

Laurie Hall at the University of Camford aspires to have the best dining facilities in the university but the President is concerned about smells. Martin, an Environmental Ergonomist and Fellow of the college, conducted an environmental ergonomics air quality survey to investigate the problem (Chapter 20). Four groups of five males and five females, from outside of the college, judged the dining hall on immediate entry and after some time in the hall (two groups while taking lunch; one while at formal evening dinner; and one after breakfast but not dining). Martin attended all sessions and made observations and took measures of carbon dioxide levels. He also assessed the thermal environment as the air temperature was known to interact with the perception of air quality.

Thermal comfort assessments predicted a mean sensation of neutral. No significant differences were found between male and female judges ($p < 0.05$). Mean

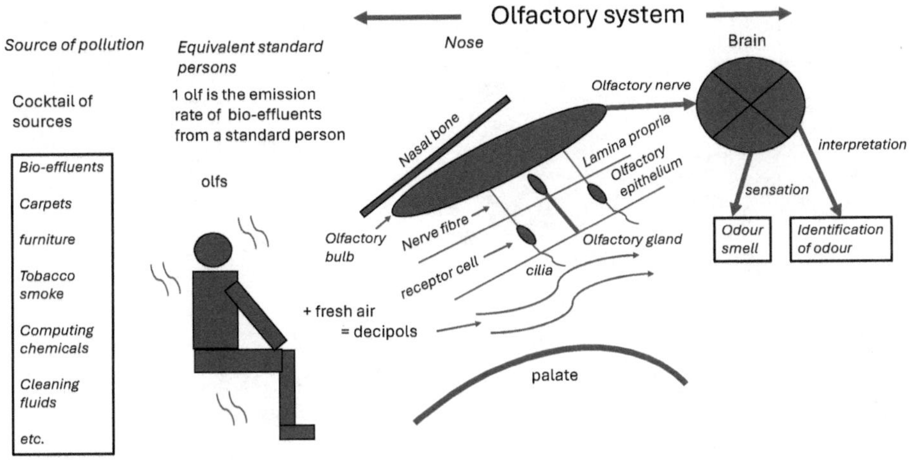

FIGURE P6 Schematic representation of the human olfactory system. Pollution sources are converted into equivalent source strength of standard persons (olfs) and with the addition of fresh air the total mixed polluted air (decipol) enters the nose (and mouth) some of which is perceived as odours and smells.

subjective responses rated the smells as strong and not fresh and most judges rated the dining room air quality as unacceptable because of cooking smells. The dissatisfaction rate was 44% providing an unacceptable 5 decipols of pollution. As a result of the survey, the ventilation system in the kitchens was redesigned and exhausts redirected. Students had not been informed of the survey and a follow-up set of questions six months later, in the subsequent student experience questionnaire, indicated that there were no air quality issues in the dining room and that the dining experience was of high quality.

19 Air Quality

AIR QUALITY

Fresh air consists of Nitrogen and Oxygen with carbon dioxide, argon and other gases in small amounts, all at fixed proportions. It is usually found outdoors. Oxygen and carbon dioxide are important as people breathe in fresh air containing oxygen at around 21 % and carbon dioxide at .03% and breathe out expired air containing oxygen at around 16–18% and carbon dioxide at around 3–5%. People produce carbon dioxide by 'burning' food in oxygen to produce energy.

Air quality is about additions and deviations from fresh air, sometimes but not necessarily detected by odours and smells. Poor air quality can cause illness and death and is a specialist subject for toxicologists, occupational hygienists, virologists and more. It can also cause unpleasant (or pleasant) smells that are not toxic but can lead to dissatisfaction with the environment and are the domain of the environmental ergonomist. Poor air quality is a common finding in many buildings with occupants complaining of stale, stuffy and unacceptable air, even when ventilation standards have been met (Fanger et al., 1988).

AIR

It is generally considered that when created the earth had no atmosphere and that the initial atmosphere was formed by volcanic gases held close to the earth by gravity. The separation of the moon and the evolution of organic life from bacteria to plants converted a high concentration of carbon dioxide to oxygen and the evolution of animal life converted oxygen to carbon dioxide resulting in a balance that today provides an atmosphere by volume of 78.08% nitrogen; 20.94% oxygen; 0.03% carbon dioxide; and 0.95% argon and others (McIntyre, 1980). Monteith and Unsworth (1990) provide integrated values for air of molecular weight 29.00 g; density of 1.292 kg m^{-3} (at standard temperature and pressure (STP sea level); 293 K, and 101 kPa); and mass concentration of 0.975 for nitrogen; 0.300 oxygen; 0.016 argon and 0.001 kg m^{-3} carbon dioxide. Making a total of 1.292 kg m^{-3}.

It is worth noting that this is the % mixture for 'fresh outdoor air' and it exists throughout the troposphere (occupied by life). So at the top of a mountain, the air may be 'thin' and less dense than at sea level, due to lower pressure, however, the % content remains the same. It is the density that changes so that 1 m^3 of atmosphere will have less amount of air up a mountain than at sea level but at the same constituent percentages. So less oxygen per breath, often not sufficient to survive and why air craft cabins are pressurised (see Parsons, 2014).

Air also contains water vapour that creates a water vapour pressure and humidity. Changes mainly in temperature but also atmospheric pressure (caused by the weight of the atmosphere) lead to condensation of water vapour and rain, ice and snow that

DOI: 10.1201/9781003401964-25

initially led to the formation of the oceans. Evaporation then returns water vapour to the atmosphere.

The amount of water vapour in the air (provides partial vapour pressure) divided by the maximum amount of water vapour that the air can hold at that temperature (provides saturated vapour pressure) is the relative humidity usually expressed as a %. The saturated vapour pressure decreases with temperature and at a temperature termed the dew point, water vapour condenses out of the air.

Saturation is more correctly an equilibrium point where evaporation and condensation are in equilibrium. Water vapour is not dissolved in air and air does not 'hold' water vapour, it mixes with it. Mixing ratio is the ratio of mass of water vapour to mass of dry air and can be derived from a psychrometric chart. Humidity in a room is often taken as a single value but will depend upon the nature and temperature of surfaces, air velocity, air temperature, water vapour generation and more.

The wet bulb and dry bulb temperatures of a whirling hygrometer (sling psychrometer) can be used to determine the relative humidity, partial vapour pressure, dew point and more from a psychrometric chart (Ellis et al., 1972).

Humidity not only affects the air quality perceived by people but also influences the growth of organisms such as moulds that release spores into the atmosphere as well as interacting with building construction materials especially when surface temperatures are below the dew point and condensation occurs on walls and windows. Boiling food in an open pot, in a sealed residential building, leads to condensation and moulds damaging walls and ceilings and creating unhealthy environments.

Enthalpy is the total energy in a system and in humid air it is the combined energy for the dry air and mass of water vapour. When air is transferred from one humidity and temperature level to another there is a heat and mass transfer the total of which is the transfer of energy (enthalpy) (see Parsons, 2014). Fanger et al. (1988) suggest that the perception of humid air is related to its enthalpy.

Piasecki et al. (2020) review enthalpy as an indicator of indoor air quality. They evaluate the proposal that acceptability (Acc) of air is related to enthalpy, h, (kJ kg^{-1}) by Acc = ah + b (a = -0.033 and b = 1162 for clean air and Acc ranges from -1, completely unacceptable through 0, to +1 completely acceptable). The predicted percentage of dissatisfied is given by PDAcc = 100/(1+(1/exp(-4.28Acc + 0.42)), (Fang et al., 1998) later modified by Toftum and Fanger (1999) to PD = 100/(1+(exp(-3.58 + 0.18(30-ta) + 0.14 (42.5 – 0.01 Pa)). (for h between 50 and 90 kJ kg^{-1}). They conclude that warm humid environments provide more dissatisfaction than often predicted by existing methods and that air enthalpy is a good predictor of responses to indoor air quality when considering temperature and humidity.

As well as the constituents that make up 'fresh air' and any water vapour content, there are additional constituents that add to the air from the surrounding environment. Together these make up the environmental air quality. They can be in gaseous, solid or liquid form.

Aerosols are disperse systems of liquid or solid particles suspended in a gas. Vapours are similar to gases but are often also found in liquid form. Humidity levels interact with other components in the air to influence air quality. Soluble chemicals can interact with humidity levels and dissolve in water vapour to form solutions that may be harmful and unpleasant. For example, water vapour combines with sulphur dioxide to form sulfuric aerosols, salts and acids.

People are immersed in the air that surrounds them and it interacts with their clothing, skin, eyes and respiratory system and breathing from mouth and nose to lungs. Depending upon how much is present in the air, additional components can cause pleasant sensations, smells, discomfort, illness and death. An old adage from occupational hygiene is that 'there is no such thing as poison – only dose'. In occupational hygiene applications it is guidance on what constitutes a safe dose that is of great importance (Harrington and Gardiner, 1995; ACGIH, 2024).

UNITS OF CONCENTRATION

The universal unit of concentration is the mass per unit volume of the atmosphere (mg m^{-3}) however gases and vapours are often described as partial volumes occupied (parts per million, ppm; or parts per billion, ppb). Vincent and Brosseau (1995). They are sometimes represented in practical terms for example, as surface area of particulate per unit volume or number of particles (e.g. asbestos fibres) per unit volume.

If the rate of emission of a pollutant is r m^3 s^{-1} and the ventilation air flow rate is Q m^3 s^{-1}, then the concentration of the substance is r/Q m^3 m^{-3} = r/Q × 10^6 parts per million (ppm) or r/Q × 10^9 parts per billion (ppb). Harrington and Gardiner (1995).

OTHER CONSTITUENTS OF AIR

The American Conference of Governmental Industrial Hygienists publish an annual booklet, easily carried in a pocket by occupational hygienists, that provides guidance on limit values for the health and well being of workers, for an extensive range of airborne substances, biological indicators and physical agents (ACGIH, 2024). These are termed Threshold Limit Values (TLVs) and biological exposure indices (BEI). Around 700 chemical substances and 50 biological exposure indices are considered as well as physical agents including acoustic, electromagnetic radiation, lifting and handling, vibration and thermal (heat and cold) environmental components.

TLVs are airborne concentrations of chemical substances below which nearly all workers can be exposed, day after day, over a lifetime of work, without adverse health effects. The threshold limit value-time weighted average (TLV-TWA) is the concentration for an 8 hour day and 40 hour working week to which workers can be exposed without adverse effect. The threshold limit value-short term exposure limit (TLV-STEL) is a fifteen-minute exposure that should not be exceeded during the work day. The threshold limit value-ceiling (TLV-C) is a concentration that should not be exceeded. A parallel system is used when considering physical agents.

Biological monitoring is usually determined using a specimen of urine, blood or expired air. Biological exposure indices (BEIs) indicate a concentration in the specimen, below which workers should not experience adverse effects on health.

McIntyre (1980) considers ventilation and air quality and notes that almost all air contains organic compounds. He considers carbon dioxide; carbon monoxide; 0zone; tobacco smoke; formaldehyde; radon; air ionisation; odours; humidity; particulates including dust as well as allergies. He notes that all limits for air quality are approximations due to the difficulty in determining cause and effect in epidemiological studies. He also presents ASHRAE standard 62–73 allowable contaminant concentration for ventilation air as annual average and short-term average respectively, in micrograms per cubic metre ($\mu g\ m^{-3}$) as follows. For particulates; 60 and 150 in 24 hours; sulphur oxides as 80 and 400 in 24 hours; carbon monoxide, 20,000 and 30,000 for 8 hours; oxidant, 100 and 500 in 1 hour; hydrocarbons (not methane), 1800 and 4000 in 3 hours; and Nitrogen oxides, 200 and 500 in 24 hours. For all other contaminants it is recommended not to exceed 0.1 of TLV values presented by the ACGIH who review limits every year.

CARBON DIOXIDE

Carbon dioxide (CO_2) level in an occupied space is the most important objective measure for the environmental ergonomics air quality survey. That is because people breath in air at 21% oxygen and 0.03% Carbon dioxide (at environmental temperature and humidity) and breath out expired air (at around 36 °C and 100% humidity) containing around 17% Oxygen and 4% Carbon Dioxide. So CO_2 levels will build up if ventilation with fresh air is inadequate.

This is important as CO_2 levels can cause headaches above 0.5% and lead to increasing illness and eventually collapse and death above 10% concentration in the air (the body can also emit other bio-effluents from all body orifices including sweat glands, that cause odours and smells that often dwell in clothing).

Measuring CO_2 levels are also a surrogate method for assessing ventilation rates and hence an indicator of the possible pollution in general caused by pollutants other than CO_2. This is now considered important as Fanger et al. (1988) and others have demonstrated that, contrary to previous assumptions, people are not necessarily the dominant cause of pollution in modern buildings.

Body emissions are related to a person's activity level as this determines their breathing and metabolic rate (energy production). McIntyre (1980) suggests that the rate of CO_2 output from a person in litres per second is given as $q = 4 \times 10^{-5}\ M\ A_{DU}\ ls^{-1}$ where A_{DU} is an estimate of body surface area (from weight and height) and M is metabolic rate in $W\ m^{-2}$ (estimated as 58 $W\ m^{-2}$ for a sitting person; 150 $W\ m^{-2}$ standing light work; 250 $W\ m^{-2}$ for heavy work, higher for sports and so on.

So ten average sized males sitting relaxed in a room ($A_{DU} = 1.8\ m^2$) would each have a metabolic rate of $58 \times 1.8 = 104.4$ W and produce $4 \times 104.4\ W \times 10^{-5} \times 3600\ s = 15.03$ litres of CO_2 each per hour and hence 150 litres of CO_2 per hour for ten men. As 1 cubic metre is 1000 litres, it would mean that in a sealed room of 10 m^3 (10000 litres), after 1 hour the percentage of CO_2 would be 1.5% and it would reach dangerous levels within 7 hours. In fact the calculation is more complex as the % of carbon dioxide in inspired air would also increase with time, not to mention the bio-effluents. McIntyre (1980) presents equations for the increase in contamination with time related to ventilation properties.

McIntyre (1980) describes the effects of increasing CO_2 levels in inspired air as beginning with 1% to 3% providing an increased rate and depth of breathing leading to discomfort and severe headaches up to 5%. A 30 minute exposure at 5% leads to intoxication and mental depression and at 10% consciousness is lost after a few minutes.

Over long periods (days) of exposure Borum et al. (1954) found acidosis and the ACGIH TLV average over a 40 hour week is 0.5% CO_2. So 0.5% CO_2 is often taken as an upper limit when assessing environments and designing ventilation systems, although when considering that CO_2 levels are an indicator of the concentration of other contaminants, less than 0.1% is often taken as good practice. For light activity it is often considered that CO_2 is produced at 0.005 l s^{-1} per person, but it should also be remembered that combustion of fuels such as natural gas, kerosene and liquid petroleum gas (LPG) all produce carbon dioxide.

The ventilation required is independent of the volume of a room and to maintain a CO_2 value of 0.5% an outdoor air supply of Q = 0.015 M litres per second per person is required, and for a normal activity (M = 65 Wm^{-2}) that is around 1 ls^{-1} per person (McIntyre (1980).

OTHER COMMON CONTAMINANTS

Carbon monoxide is a colourless odourless gas that is produced by incomplete combustion of fuels from traffic to heating. It replaces oxygen by attaching to haemoglobin in blood and quickly produces severe symptoms from headaches to vomiting to death within one hour depending upon concentrations. It is the most common cause of death and accounts for more deaths in the world than all other gases put together.

Ozone (O_3) is a reactive allotrope of oxygen. It has a fresh smell and can cause eye irritations, however it is toxic because of its chemical activity. It is usually found outdoors and is a constituent of smog. It is not usually found indoors and decays quickly when it is. I was surprised when I attended the international conference on environmental ergonomics in San Diego, when I complained that my non-smoking room smelled of smoke, an ozone machine was sent to my room. It seemed to do the trick, at least I did not complain further!

Tobacco smoke is not often found indoors in many countries due to legislation, although the environmental ergonomist may wish to check the building air intake system, including doors and windows, to ensure that outdoor smoke is not taken into the building. Ventilation rate requirements are often high for rooms where smoking occurs because of the increased recognition of the danger of tobacco smoke to health (e.g. >20 m^3 per cigarette). It is an example where some people might regard the smoke as pleasant (e.g. cigar smoke) but it is a danger to health. Ventilation rates for vaping systems used to replace smoking can be derived but there is a lack of data on their effects on health, if any.

HUMAN INTERACTION WITH THE ATMOSPHERE

People can sense the atmosphere in which they are immersed and judge it accordingly, usually through a sensory system that detects a subset of the constituents of the air that it interprets as odours, and if the odour is unpleasant, a smell.

The (clothed) human body is surrounded by the air in the environment that is in contact with the skin, eyes and the nose and mouth through breathing. The skin is underestimated in its interaction with the air and, particularly when sweating, it can absorb substances, through pores and sweat glands, into the blood stream. It is often inadequate therefore to protect only the mouth, eyes and nose (with facemasks) when exposed to dangerous atmospheres (chemical and biological).

Full nuclear, biological and chemical protective clothing is often used by military (and emergency services) personnel (with activated (increased surface area) charcoal in material to absorb contaminants) and can be effective but cause loss in human performance at tasks as well as sweating. So protective clothing is a full systems operation, not just a matter of, but including, donning and doffing. Eyes can be particularly vulnerable to some chemicals and may need to be protected.

'Although air quality standards are set in practice by physical terms of contaminants, in practice it is the level of odours by which the air quality in a space is judged'. McIntyre (1980). The environmental ergonomist is concerned with the health, comfort and performance of people who occupy a space and in terms of air quality, this is mainly restricted to subjective measurements of odours and smells as well as measuring levels of carbon dioxide.

A follow-up assessment would probably be conducted by specialists in toxicology, ventilation and experts in the particular chemical or chemicals involved. Many constituents of the air are odourless and often important for health but they are not usually the concern of the environmental ergonomist, other than emphasising their possible presence in any report.

PERCEPTION OF AIR QUALITY

Air is breathed in through the nose and mouth and chemical reactions in specialised cells, (chemoreceptors) mainly in the nose, detect odours. These cells are replaced as they wear out. Taste and smell are related and if the nose is blocked, taste is reduced.

The mouth detects taste and the texture, spiciness, temperature and odour of food. The sides of the tongue and back of the mouth contain taste buds from front to back respectively for sweet, salty, sour, bitter and savoury (umami) sensations. Food dissolved in saliva enters taste pores and stimulates taste hairs. Many tastes can be determined by interpretation of the stimulation of a combination of the basic tastes, complemented by odours.

Air enters the nose and some of its constituents react with a small area (2.5 cm^2) in the roof of the nasal cavity in each nostril, called the olfactory epithelium. This area contains chemoreceptors (an order of magnitude more sensitive than those for taste) and with interpretation a person can detect over 10,000 different odours.

Receptor cells are olfactory neurons that extend through a mucus like fluid on the olfactory epithelium, that dissolves odours and are detected by cilia that send signals to the olfactory nerve and via the olfactory bulb to the olfactory centre of the brain for interpretation.

Two points are relevant. Of great importance is that the chemoreceptor cells can saturate and people can adapt to a 'smelly' environment. As a consequence a person entering a room may perceive it as smelly but the occupants who have been in the room for an hour or so will not perceive the smells and may regard it as acceptable.

Assessment of air quality may then have to be made by people from a fresh air environment providing subjective responses on directly entering the room.

The second point is that the sense of smell may be lost due to some bacterial or viral illnesses (e.g. coronavirus – COVID). So much so that it may become a recognised symptom of the disease and even used as an identification and filtering system for identifying those fit for work and others who may pass on the illness to fellow workers, customers, clients, etc.

PERCEIVED AIR QUALITY

A colleague of mine told the story of a disgruntled employee (some say it was his father) who, late one Friday afternoon, nailed a kipper (smoked herring) to the underside of his desk as he left for retirement. When workers arrived back to work on Monday, there was a distinct and unpleasant fishy smell in the office and people refused to work there. When the kipper was eventually located (with the assistance of dogs, Daisy and Lulu) it had a smell of rotten fish and the whole area had to be cleaned and flooded with fresh air and hygienic cleaning odours. The point of the story is that people can detect and identify odours, that undesirable odours can be described with the generic term 'smelly' with sufficient intensity to be unacceptable, that fresh air is desirable (but is odourless) and that there are odours that can be identified as desirable and even pleasant. There also appears to be some agreement about what are good and bad odours.

Air quality is perceived by a person during the inhalation phase of breathing through the mouth and nose. There is a rich supply of semantic terms, world-wide, but there is no comprehensive understanding of how they are made up from environmental parameters, particularly when there is a mixture (cocktail) of constituents in the air.

Attempts at producing an artificial nose to simulate human response to odours with reliable units have been made but for the perception of air quality, subjective methods are required. An olfactometer is a device that allows different gases to be inhaled and odour strength and perceived smell documented but for the environmental ergonomist, it is the subjective assessment of the actual air under consideration that provides the main assessment of the perceived air quality in an environment. Chrenko (1974), from Bedford (1948) provides a history.

UNITS FOR THE MEASUREMENT OF AIR QUALITY

The constituents of air can be detected by people (odours) or not and in turn they can be harmful or not. Environmental ergonomics is usually concerned with odours that are not toxic (considered by specialist experts) but can affect the comfort, satisfaction, performance and well-being of people.

The thermal environment and air quality are closely linked when considering human response to the environment. Opening windows for fresh air or a breeze will affect both air quality and the thermal environment (not to mention the acoustic environment).

Increased humidity levels can reduce levels of odour perception. Keuhner (1956). It is widely recognised that cold air is more refreshing than warm air with a higher air quality for the same constituents. Sir Lenard Hill (1914) (see Hill et al., 1916) added to this in sealed thermal chamber experiments, showing that for the same constituents of air, fans provided relief from discomfort. It is therefore sensible to conduct a thermal environment (comfort) survey to ensure an acceptable thermal environment before making an environmental ergonomics air quality survey.

It has been recognised for hundreds of years that 'bad air' is related to discomfort and illness, that pleasant gases can mask unpleasant smells and so on. Reid (1844) observed food consumption at Royal Society dinners in Edinburgh and concluded that increased air quality improved appetite (also judged by the consumption of wine). As the scientific method developed, attempts have been made to relate the make-up of air to the effects on a person. All methods have concluded that the nature of smell is best determined subjectively.

A particular subjective method that is useful in assessing sensations is to use an expert panel. Wine tasting is an excellent example where the 'taste and nose' of a panel of trained experts can assess the quality and characteristics of wine in a reliable (repeatable) way. Similar methods are used to assess perfumes and other odours and attempts have been made to relate the outcomes to scales of odour type and intensity. A panel may not need experts to make judgements but simply the use of their noses as superior measuring instruments than any non-organic devices. While the use of panels will provide general results for a population, subjective assessment using the actual occupants of a room under investigation will give relevant results in context.

Ventilation is the supply of fresh air to a space to meet requirements. McIntyre (1980) notes that ventilation was initially provided as an attempt to control diseases but nowadays it is mainly thought of as the provision of air for respiration; the dilution of odour; the removal of excess moisture; the dilution of contaminants; the provision of air for combustion and for temperature control. It may be that in the future it may resume its role as a control for disease and in particular the transfer of bacteria and viruses in attempt to reduce infection rates.

Environmental ergonomics is usually confined to the assessment of odours, typically in indoor environments. It is well known that on initially entering a room odours are easily detected, but that people who have been in the environment for some time adapt and cease to detect the odours.

Yaglou et al. (1936) attempted to improve existing ventilation standards. They defined the absence of disagreeable odours as a satisfactory ventilation rate. Importantly they took two bases for determining fresh air requirements. They were the primary impression on entering a room from relatively fresh clean air and the impressions of the occupants in the room.

They found that the intensity of odour in a room varied with the number and personal hygiene habits of occupants as well as the fresh air supply per person. They suggested, for sedentary adults of average economic status, ventilation rates required varied from 25 cu.ft/min/person for a room with an occupied density of 100 cu.ft per person to 16 cu.ft/ min/person for a density of 200 cu.ft per person and 7 cu.ft/min/ person for a density of 500 cu.ft/person. For a density of 200 cu.ft/person, Labourers required 23 cu.ft/min/person and school children from 18 to 38 cu.ft/min/person depending upon social class. For most ordinary conditions a ventilation rate of 17 to 20 cu.ft/min/person is suggested (Note that 100 cu.ft = 2.83 m^3).

Two seminal papers were produced by Fanger (1988) and Fanger et al. (1988). They were stimulated by the emergence of what were called sick buildings, where people were complaining about environments in buildings in which they worked as well as refusing to work in them. This was termed sick building syndrome (headaches, nausea, lethargy and more) and appeared world-wide during the period when there was an increasing prevalence of the use of solvents, modern building materials, printers and materials associated with computer work stations and visual display units (also old carpets, rooms in need of refurbishment and more).

Two new units were developed. The olf is related to the emission of odours and the decipol related to perceived intensity. It was suggested that people may not be the sole cause of pollution in buildings but that any source could be related to the equivalent emission provided by a person.

The olf unit was defined as the bio-effluent emission of a standard person who is in thermal neutrality; sitting at rest in light clothing (1 met, 0.6 clo); 1.8 m^2 body surface area; with a hygiene level of 0.7 baths per day (and using deodorant); and changes underwear daily. All other emission sources were then converted to the equivalent emissions of number of standard persons and hence number of olfs. So one olf is the emission rate of air pollutants (bio-effluents) from a standard person. Any other pollution source is quantified by the number of standard persons (olfs) required to cause the same dissatisfaction as the actual pollution source (Fanger et al., 1988). The equivalence is therefore in terms of the same level of dissatisfaction; a subjective term.

The perceived air pollution is that concentration of human bio-effluents that would cause the same dissatisfaction as the actual air pollutants. The pollution caused by the emission of odours from the equivalent number of standard persons (olfs) depends upon the ventilation rate (of fresh air). The decipol is the pollution caused by one standard person (1 olf) ventilated by 10 litres per second of unpolluted air. It is emphasised by Fanger (1988) that the olf and the decipol are applicable to air pollution perceived by people indoors and outdoors.

The olf was derived in two auditoria in Denmark occupied by 1000 people in thermal neutrality, for at least 20 minutes, and the air quality was judged by 168 young men and women just after entering the space. They were asked to say whether the air quality was acceptable or not in the context of imagining themselves having to enter the space frequently during their work.

Analysis led to the percentage of dissatisfied judges (PD%) as a function of the ventilation rate per olf (q ls^{-1} olf^{-1}) predicted as $PD = 395 \exp(-1.83\, q^{0.25})$ and for $q<0.32$ ls^{-1} olf^{-1}, PD=100%. As 1 decipol is 0.1 olf per l/s, $PD = 395 \exp(-3.25\, C^{-0.25})$ where C is the perceived air pollution in decipols.

In short if we determine the dissatisfaction rate, we can calculate the pollution (decipol) level and if we know the ventilation rate we can calculate the emission level (olfs). If we determine the percentage dissatisfied with no people present and ventilation rate on or off we can determine the olf value for the contaminants in the air and ventilation system, not attributable to people.

Fanger et al. (1988) determined the olf values for 15 offices and 5 assembly halls ('randomly' selected buildings with no known air quality problems) for conditions of no people, ventilation off (at least 12 hours before assessment); no people, ventilation on (for several hours before judgement); and people (some smoking) with ventilation on. A total of 54 judges (27 male, 27 female, aged 18 to 30 yrs, including 17 smokers - selected from advertisement) visited each room on three separate days (in balanced order of conditions).

Before entering the room and outdoors, each judge rated outdoor air in terms of odour strength (no odour to overpowering odour); acceptability (acceptable or not acceptable); and freshness, (very fresh to stuffy). On entering the room they used the same scales to judge the indoor air (imagining that they should be exposed to such air during their daily work). At the same time, ventilation air change and air flow was measured using a tracer gas technique (Dietz et al., 1986), and carbon dioxide, carbon monoxide, volatile organic compounds and particulates were measured in the occupied space 1.1 m above the floor.

The results showed that outdoor air could be considered as fresh air and at around 8 °C and 90% humidity with light wind. Indoor air was around 22 °C and 42 % humidity. Maximum CO_2 values did not exceed 0.082% above outdoor levels.

> For each occupant in the15 offices there were on average 6–7 olfs from other pollution sources; 1–2 olfs were situated in the materials in the space; 3 olfs in the ventilation system; and 2 olfs were caused by tobacco smoking. The ventilation rate was 25 l/s per occupant, but due to the extensive other pollution sources only 4 l/s per olf.

30% of the judges found the air quality in the offices unacceptable (Table 19.1).

There were large individual differences between buildings in terms of judged air quality but they could not be predicted from objective measures. Important findings were that occupants were not the primary source of pollution and therefore ventilation guidelines, based only on that, will be limited. Also that the ventilation system itself may be an important source of pollution partly defeating its purpose (Table 19.2).

It was recommended that to aid selection of materials, olf values for different materials should be determined under controlled conditions in the laboratory (See Yegeneh et al., 2006). Fanger (1989) proposed a new comfort equation for indoor air quality as $C_i = 112\ (\log_e(PD) - 5.98)^{-4}$ where C_i is the perceived indoor air quality in decipols and PD is the percentage of dissatisfied (%). A more general comfort equation involving olfs and ventilation rate is provided but the environmental ergonomist may estimate the perceived air intensity in decipols from the percentage of dissatisfied from subjective judgement.

ASHRAE (1989) specifies that the air quality should satisfy 80 % of occupants, hence 20% dissatisfied or 1.4 decipols (assume steady state conditions and complete mixing of the air). To obtain an indoor air quality of 1.4 decipol a ventilation rate of 5 l/s per m^2 is required for average existing office buildings (Fanger, 1989) who also suggests that for low olf buildings the ventilation required is 1.4 l/s per m^2 (of floor area).

TABLE 19.1
Mean values for air quality in 15 offices

Air change (l/h)		CO_2%	CO ppm	Part µg/ m3	VOC mg/m3	Odour intensity	Dissatisfied %	Freshness rating
Ventilation off, no occupants	0.4	0.006	0.0	-21	0.15	1.8	34	-0.5
Ventilation on, no occupants	2.0	0.002	0.0	-21	0.11	1.5	29	-0.4
Ventilation on, Occupants	2.1	0.032	0.3	40	0.20	1.6	34	-0.5

(Data from Fanger et al., 1988). CO_2, CO and particulate values are levels above outside air values. Odour intensity of 1 is 'slight odour' and 2, 'moderate odour'. Freshness is 0 for 'neutral' and -1 for slightly stuffy.

TABLE 19.2
Mean pollution levels (decipols), ventilation rates and olf (equivalent standard persons) values showing main source of pollution is not occupants

	Floor area m²	Outdoor air supply (l/s)	Pollution sources olf/m² floor	Ventilation rate		Perceived air pollution (decipol)
				l/s occup	l/s olf	
Mean 15 Offices	230	398	0.60	25.0	4.4	3.3
Mean 5 Assembly halls	130	618	0.74	97.5	17.8	1.3

(From Fanger et al., 1988)

ANSI/ASHRAE STANDARD 62.1-2022: VENTILATION AND ACCEPTABLE INDOOR AIR QUALITY

ANSI/ASHRAE (2022) define acceptable indoor air quality (IAQ) as air in which there are no known contaminants at harmful concentrations, …and with which…80% or more of people exposed do not express dissatisfaction. It is a comprehensive document and it is part 1 of a two part document. Part 2, is concerned with residential buildings and non-transient populations. The definition provided allows the environmental ergonomist to compare subjective results of an air quality survey with accepted national and international standards.

Fanger (1988; 1989) compared the olf as analogous to the lumen for light and the watt for noise emission and the lux for light and decibel for noise perception. He also provides an 'air balance (pollution balance)' comfort equation for air quality analogous to the heat balance equation for assessing the thermal environment. That is the outdoor air supplied to the space (Q l/s) must balance with the pollution sources in the space (G olfs) so Q = 10 G/ (Ci – Co) where Ci and Co are the perceived indoor and outdoor air quality, respectively, in decipols.

Using his comfort equation, Fanger (1989) estimates olf values for offices with one occupant per 10 m^2 of floor area as between 0.2 to 0.7 olfs per square metre with ventilation requirements of 1.4 ls^{-1} per m^2. This can be compared with 0.8 ls^{-1} per m^2 for ASHRAE 62-89; 0.4 ls^{-1} per m^2 for Nordic guidelines (NKB 1981) and 1.4 ls^{-1} per m^2 for DIN 1946 (1983) assuming no smoking in offices.

A note on the method used is that the initial response is taken as the measure of acceptability and dissatisfaction. This is probably the most important measure and in specifying ventilation required, provides a safe margin to ensure that air quality is maintained throughout the occupation of the environment. It does not however take account of the adaptation of occupants to odours over time, where, although complex, if managed carefully may offer an opportunity for savings in expense and energy consumption.

A further point is that on making a judgement initially, the transient effect of a changing environment (from cool, (breezy) and fresh outside) to a perceived warm (even at steady state thermal neutrality) environment, even at the same level of air quality, may not be perceived of as fresh as outside. It would however be a realistic simulation of what occurs in practice.

Chapter 19 provides a case study describing the environmental assessment of air quality in a university dining room. It uses the principles and methods described in the present chapter.

FURTHER READING

McIntyre, D. A., 1980, Indoor climate, Applied Science, London, ISBN 0-85334-868-5

Fanger, P. O., 1988, Introduction of the olf and decipol units to quantify air pollution perceived by humans indoors and outdoors, Energy and Buildings, 12, 1988, pp 1–6.

STANDARDS

ANSI/ASHRAE Standard 62.1, 2022, Ventilation and acceptable indoor air quality, ASHRAE, Atlanta, USA

20 Case Study
Environmental Ergonomics Assessment of Air Quality in a University Dining Hall

THE DINING HALL

Laurie Hall, a college at the University of Camford with around 1000 undergraduate and postgraduate, mainly mature, students, has recently extended the old college and moved the dining room, along with various works of art, to overlook the cricket field and provide a pleasant and prestigious space for the college.

Martin, a professor at the Institute of building ergonomics was a Fellow of the college with dining rights. At one formal dinner on the top table he sat next to the President of the college. The president said that he was content with the new dining hall but that he thought that there were unpleasant cooking odours and that although no-one had complained, he was concerned that the dining experience could be much improved by their elimination.

Martin suggested that a redesign of the ventilation system for the adjacent kitchen would probably solve the problem, but he would be happy to conduct an environmental ergonomics air quality survey to determine the extent of the problem before the expense of a redesign was embarked upon. The President gratefully accepted the offer and referred Martin to the Bursar who was sitting further down the table.

THE MEETING

Over Port and post dinner drinks, Martin and the Bursar agreed to the details of the survey. They met on Monday morning for coffee with the head of the kitchen who explained that all meals were buffet style, apart from formal dinners. Baking started early in the morning leading up to breakfast for 300 residential students between 7.0 am and 9.0 am. At 10.0 am lunch was prepared and lunch was served between 12.0 noon and 2.0 pm. Dinner was prepared from 3.30 pm and served from 5.30 pm until 7.0 pm. All colleges aspired to a reputation for fine dining and the chef was proud of the dining experience provided by Laurie Hall. His view was that dining rooms were supposed to smell of food, it was part of the experience.

It took the president to re-assure the chef (of some note), in diplomatic terms, that such a talented team, providing outstanding fare, deserved a dining room of the highest quality and that he, the president had as part of his strategy for the college, an aspiration to provide the best dining experience in the university.

DOI: 10.1201/9781003401964-26

Martin explained that in the first instance he would use subjective judgements to assess whether the air quality in the dining room was acceptable or not in the context of dining. If a problem was found he was sure that a bespoke ventilation system would solve the problem.

The Bursar decided that the nature of the investigation would not be advertised but if asked, be presented to diners, as an 'icing on the cake' exercise and suggest that the survey was to ensure that the college had one of the very best dining halls in the university.

AIM

It was agreed that the specific aim of the project was to determine whether the dining hall air quality was commensurate with satisfied diners in a top-quality dining room environment.

THE INVESTIGATION

Three groups of five male and five female researchers from Matin's Institute (not at the college), aged between 18 and 30 years, were selected as judges to dine at the college (no shortage of volunteers). The judges were of an international profile, similar to, but not matched to, those of the college.

A typical term-time week was chosen. The first group attended lunch on Tuesday, the second group attended dinner on Wednesday and the third a formal dinner (in formal dress) on Thursday evening from 6.0 to 8.0 pm. A fourth group of judges did not dine but assessed the dining room on Friday morning at 9.45 am.

Two forms of judgements were made for each assessment using identical scales. All were made discreetly on scales on mobile telephones with 'comment boxes' for additional information. One judgement was made on immediately entering the dining room from fresh air outside (down a 10 m corridor to the dining room entrance), the second was made after the dining experience but still in the room. Judges were instructed to make their own judgements and not to discuss the survey with colleagues or anyone else.

The dining room was on the ground floor (with terrace above) and had dimensions 35 m × 20 m with windows (and emergency door) on the long 'wall' towards the west (overlooking the cricket pitch, with nets to stop hits to the boundary for 4 and 6) and solid walls with paintings towards the east, with entrance and exit doors. The kitchen entrance and exit was through separate swing doors on the south wall and the top table was on a stage with solid wall behind supporting a large feature abstract painting of some note. Martin was supplied with a blueprint of the building including dimensions and services, as well as layout of the kitchen and equipment.

After one hour into each assessment period, Martin used a whirling hygrometer to determine air temperature and humidity in the dining room and with estimation of other factors (clothing, etc.) confirmed that the diners would be around thermal neutrality, while essentially sitting at rest. He also counted the number of diners (at any time) over the period of the meal and observed seating locations (selected by the diners). He monitored CO_2 levels at the centre of the room where he sat.

On Tuesday and Wednesday, judges waited outside and away from the dining room for 30 minutes in around 10 °C air temperature and some wind. They entered the dining room in male/female pairs at ten-minute intervals from the opening of the dining room, along with the regular diners, and sat in pairs at designated areas to provide a spread across the room.

Immediately on entering the dining room they rated whether the air quality was acceptable or not in the context of it being a dining room. They then rated odour strength (no odour; slight odour; moderate odour; strong odour; very strong odour; overpowering odour), freshness (not fresh; slightly fresh; fresh; very fresh; extremely fresh), and whether the air quality environment was welcoming or not in the context of the dining room. A rating of how do YOU feel NOW; on the scale; hot, warm, slightly warm, neutral, slightly cool, cool, cold was also provided.

The scales were quickly completed on the mobile telephones with little interference to activity and were completed again after the meal was finished, in the same location and before coffee. No alcohol was consumed.

As it was a formal meal on the Thursday evening, all diners took their positions at the same time and with silver service, a similar protocol was used for the 10 judges. With two judges on the top table. On Friday morning the ten judges used an identical protocol but did not dine and provided responses only on entering the hall, and not after a period of time. So, 40 judges in all, each providing judgements immediately on entering the room and the 30 judge diners, also providing judgements after a period of more than one hour of exposure (i.e. 7 sets of 10 judgements). Judges were instructed not to discuss their experiences with others, at the Institute or elsewhere, for at least one week.

The mobile telephone responses were sent directly to a central computer and analysed using bespoke software with presentation of raw data as well as integrated results. In the first instance the data for the four groups (days) were analysed for judgements made on first entering the room.

Non-parametric statistics were used for analysis, along with graphs and tables of data. A Mann-Whitney U test over all 40 judgements, showed no statistical difference between male and female responses ($p < 0.05$). Odour strength varied from slight odour to very strong odour (with a modal rating of strong) and all of the responses rated not fresh. Thermal responses on first entering the room (from a cool outside) varied from neutral to warm. Comments all identified the odours as cooking smells, 5 judges said pleasant and welcoming cooking smells. 20 of the 30 judge diners on entering the room, rated it as unacceptable and 3 of the judges on the Friday at 9.45 am, rated it as unacceptable (see Table 20.1).

Judgements after at least one hour in the dining room showed a reduced modal odour strength to slightly strong. It was still rated as not fresh and 60% of judgments found the environment not welcoming especially on entering the room from outside.

Measurements suggested that the room maintained a thermoneutral air temperature (PMV = 0.3, ISO 7730, 2024) of 23 °C. The diners found the room warm after entering from outside which was not surprising as indoor clothing was worn outside and inside, but it settled to slightly warm which is probably related to activity. It was recommended not to let the air temperature drift upwards during dining and increased occupation.

TABLE 20.1
Summary responses of 10×4=40 judges and measurements(number, mode, value) e.g. on tuesday entering the dining room, 7 of 10 judges found it not acceptable; most found the odour strong; not fresh and warm. 3 judges found it welcoming.

Not Acceptable		Odour	Freshness	welcoming	thermal air	temp°C	Humidity%	CO_2%
Tues enter	7	strong	not	3	warm	22	50	.03
Tues >1hr	3	slight	not	6	sl.warm	23	58	.04
Wed enter	8	strong	not	1	warm	22	53	.03
Wed >1hr	3	slight	not	6	sl.warm	23	60	.05
Thurs enter	5	strong	not	3	warm	22	60	.03
Thurs >1hr	2	slight	not	6	sl.warm	23	62	.06
Fri Enter	3	slight	not	3	sl.warm	22	50	.03

The limited objective measures taken showed that the heating and ventilation system for the dining room was capable of providing thermal comfort and that carbon dioxide levels did not 'build up' and could be controlled by current systems.

The specific finding of the study, related to its aim, was that significant numbers of diners found the air quality unacceptable, and therefore that at least 44% of users of the dining hall would be dissatisfied.

The emission of odours to the dining room appeared to come from the kitchen and the pollution caused will depend upon how much the odours are diluted by ventilation using fresh air. The olf is the emission of one standard person. The decipol unit is a measure of pollution caused by one standard person ventilated by 10 l/s of unpolluted air.

Traditionally smells in an environment were substantially caused by people, hence the nature of the standard used for equivalence. However, It is possible to relate this to any contaminant including cooking smells and refer to the number of decipols that would be acceptable.

The dissatisfaction level is related to the level of pollution so the decipol level can be estimated from the percentage of dissatisfaction. An estimated pollution level of a standard environment would expect less than 20% of dissatisfied and 1.4 decipol. For a gold standard 10% dissatisfied (<1 decipol) would be acceptable. For a dissatisfied rating of 44% we would expect pollution of around 5 decipol. The ventilation level to reduce the pollution from 5 decipol to <1 decipol can be determined. However it is best to reduce the emission level at source by bespoke ventilation in the kitchen so that odours do not reach the dining room.

A full report with all data was sent to the Bursar with recommendations. A letter was sent to the President as follows:

Dear President

We have now completed the environmental ergonomics air quality survey of the Laurie Hall dining room. As you suspected we have found that the air quality in the dining room does not meet the high standards to which you aspire.

I have sent a full report to the Bursar, however in brief we have found significant dissatisfaction with the air quality in the environment that needs attention. I would emphasise that I am confident that there should be no concern over not meeting health and safety standards.

It would be possible to design a ventilation system for the dining room that would dilute the odours, however it is best practice to solve the problem at source. With your permission therefore I will contact Gill who is a Professor in ventilation and air quality from the Institute who is an expert in the design of bespoke (and quiet) ventilation systems. She will be able to provide a solution with costs for a ventilation system for the kitchen, and how to use it, to ensure that the Laurie Hall Dining room has gold standard air quality for the future.

If you agree then I will ask Gill to contact the Bursar in the coming weeks.

Kind regards

Martin

THE OUTCOME

The Bursar read the report in full and advised the President to go ahead with the project. Gill redesigned the new ventilation for the kitchen (with the help of Richard, a noise reduction expert) that involved vertical ventilation, expansion boxes and an exhaust system that did not re-enter the building.

The work was carried out without disruption to services, including a normal 'down time' period over the summer when maintenance was already scheduled (students dined in another college in a quid pro quo arrangement).

After the dining hall had been in operation for a few months (with a new cohort of 'freshmen' students) an assessment was made indirectly through the annual student experience questionnaire that was conducted online by all students in the college. The questions were embedded in a section concerned with the college environment in general and specifically related to the dining room. A scale of how do YOU find the air quality in the college dining room? clearly acceptable; neutral; clearly unacceptable. 78% of responses indicated clearly acceptable, 18% indicated neutral with some suggesting that they did not understand the question and 4% said unacceptable but gave no reason.

On his retirement, the Dining Hall was named after the President who commented that the roast swan had never tasted so good!

FURTHER READING

Fanger, P. O., 1988, Introduction of the olf and decipol units to quantify air pollution perceived by humans indoors and outdoors, *Energy and Buildings*, 12, pp 1–6.

STANDARDS

ANSI/ASHRAE Standard 62.1, 2022; *Ventilation and acceptable indoor air quality*, ASHRAE, Atlanta, USA.

Part VII

Environmental Ergonomics for Diverse Environments and Diverse Populations

Part 7 is in two chapters. Chapter 21 considers environmental ergonomics in diverse environments and Chapter 22 considers environmental ergonomics and special requirements for diverse populations.

Human response to the environment includes any environment people occupy or can occupy. In principle, there are an infinite number of such environments and at the time of writing there are over 8 billion people who can occupy them. In addition, they are all changing with time. Environments continuously change and not only are people different at birth but they all have increasingly different experiences after that (Figure P7).

Any categorization of individuals or environments must be for a purpose. Environmental ergonomics investigation is not a formula, it must be versatile and bring to bear methods and techniques to match the context, taking account of environmental conditions and populations, to meet objectives.

Categorization of environments is often related to location but in practical application physical (psychological, social etc.) properties will be important. Air temperature, light levels, pressure, gravity, crowding and so on will influence details of assessment methods. Categorization of people can be controversial as it brings with it inherent bias and political connotations. Dividing people into group A or B may seem sensible, but why? Even with the use of a null hypothesis, the division has already provided bias. Acknowledgement that all people are individuals is important but insufficient for design. Designing to account for the requirements of groups and

DOI: 10.1201/9781003401964-27

populations becomes a practical necessity. Inclusive design cannot always be complete. It is defined in terms of 'widest possible' and categories may be unavoidable.

Sampling may be necessary in an environmental ergonomics investigation and techniques to ensure samples are representative of both the population of interest and the environments to which they are exposed will be required.

People exist in wide range of environments on earth and it will not be long before significant numbers of people travel to space, earth's moon and other planets. Chapter 21 considers environmental ergonomics in diverse environments and special requirements for people outdoors; in vehicles; underground in, on and under water; and in air, in space, on the moon and beyond. For each environment, the effects of thermal; sound; vibration; light; and air quality environmental components are considered. It is concluded that much is known about human response to the environment and that it can be adapted to consider diverse and even novel environments. Subjective, behavioural, objective and modelling methods will apply (Figure P7). For any environmental ergonomics investigation, there is no formula for design. Methods must be adapted to the context and the aims of the investigation and that includes the challenge of assessing diverse environments.

Chapter 22 considers environmental ergonomics for diverse populations. There are over 8 billion people on earth and each person is unique. Not only is each person born unique, people become increasingly different as they have unique experiences. Ergonomics and inclusive design includes how to design environments for the widest range of people. Environments that are accessible to, and usable by, people with the widest range of abilities, within the widest range of situations, without the need for special adaptation or design. Designing for adaptive opportunity complements that philosophy. National Geographic location; age; gender; People with disabilities; and people with sensory illness and impairment are considered. Responses to environmental components: thermal; sound; vibration; light; and air quality are considered for each population.

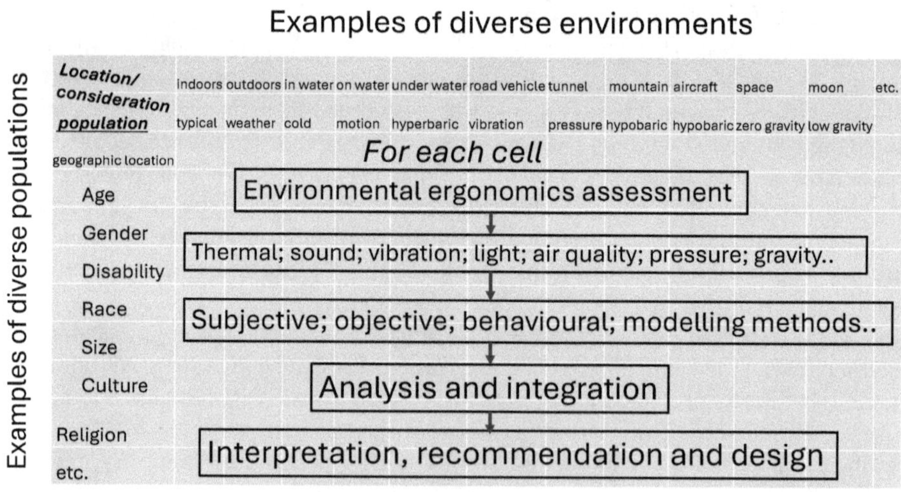

FIGURE P7 Matrix showing a wide range of combinations of diverse populations and diverse environments. Each cell represents a total integrated environment for environmental ergonomics assessment.

21 Environmental Ergonomics in Diverse Environments

DIVERSE ENVIRONMENTS

The physical environment exists outside of the body and is fundamentally determined by the dynamic relationship between the earth and its atmosphere, the moon and sun and beyond. People can exist outdoors, subjected to the weather and sometimes temporarily in (head above) water, but they often exist in micro-climates to provide protection and survival, comfort and performance when direct exposure to the physical environment is not acceptable.

Micro-climates include buildings, vehicles, space stations, inside clothing including protective clothing and more, and greatly extend the outside environments in which people can live and survive. There are therefore a large number of diverse environments relevant to environmental ergonomics assessment.

'Typical' environments may include, homes, offices, public buildings, outside recreational areas and so on, but the totality of environments in which people exist is diverse. All can influence the health, comfort and performance of people and are the domain of environmental ergonomics. Methods used may be consistent with those used for 'typical' environments described in previous chapters, but they are often considered by ergonomists and others with a specialist knowledge of the environment.

People are exposed to environments indoors and outdoors. Indoor environments isolate people from outdoors usually for protection, privacy, comfort and pleasure, as well as for control of the environment to which they are exposed. Outdoor environments are by definition, those that are not indoors so all other environments. A more functional definition maybe those environments that are dominated by outside conditions. Camping in tents for example is often referred to as an outdoor activity. Working in an open-ended car port could be an example of a transition between indoors and outdoors. A glass greenhouse is designed to be influenced by both indoor and outdoor environments. Are vehicle interiors indoors or outdoors?

If an environment occupied by a person or people is protected from the outside environment, as long as the environment is correctly designed and controlled, in theory, they can exist anywhere on earth, in space and in the universe (although on the sun would be a challenge!). We can consider typical indoor environments as those occupied by people in buildings such as houses, offices, public buildings and so on. Diverse indoor environments would require special considerations to provide optimum and acceptable conditions.

DOI: 10.1201/9781003401964-28

People have evolved to exist in physical environments on earth. It is worth noting therefore that if they travel beyond earth, physical factors such as electromagnetic radiation and varied gravity may have profound affects not yet understood, even if short-term micro-climates (in space or even buildings on the moon) may suffice.

People will be aware of the environment and context of their environment with a corresponding psychological and social disposition. Strict categorisations of physical environments are not easily defined . For convenience, consideration of environmental ergonomics in diverse environments is provided below in terms of physical environments; outdoors; in vehicles; underground; in water and underwater; in the air; in space; on the moon and on other planets. In each of those contexts, the effects of environmental components; thermal; Acoustic; Vibration; Light; and Air quality, will be considered.

ENVIRONMENTAL ERGONOMICS METHODS FOR DIVERSE ENVIRONMENTS

Even when knowledge is not complete, it is useful to adopt the philosophy that environmental ergonomics can be considered in terms of the effects of environmental components and their integrated effects. For each environmental component and its integration therefore the combined use of subjective, objective, behavioural and modelling effects remain appropriate. The emphasis on each will depend upon the context and nature of the environment.

For comfort, subjective and behavioural methods will be important, but in extreme environments objective methods may be more valid (as extreme environments can cause strain that can interfere with judgement and cognitive capacity). Modelling, including environmental indices, will always have limitations but will be particularly useful for predicting responses to extreme or proposed environments. They will be useful for predicting the likely responses of populations and for providing guidelines and limits to provide optimum environments, or to prevent unacceptable outcomes. They are particularly useful for varying parameter values and testing scenarios.

Behavioural measures are informative when subjective methods are not possible or reliable (e.g. in 'kindergarden') and are also important for all environmental components and their integration through the provision and uptake of adaptive opportunities. Generic adaptive behaviours for all environmental components include moving away from discomfort or a hazard or not entering the environment, changes in posture and orientation and using protective devices. These are not repeated below as well as the principle of always trying to reduce the hazard at source, which will always apply to stressful environments.

OUTDOOR ENVIRONMENTS

Thermal Environments

Outdoor environments are exposed to the weather, including wind and sun and often rain, snow and more. At night there is no direct solar radiation but thermal radiation can be lost to space. The principles, methods and case studies presented in Part 2 of this book are applicable to outdoor environments. In particular ISO 7243 (2017) uses the WBGT index for the assessment of heat stress and is derived from early work on army recruits, outdoors in the heat; ISO 7730 (2024) for thermal comfort has been evaluated world-wide and ISO 11079 (2007) for cold stress includes the wind-chill index and temperature for outdoor environments as well as limiting temperatures for hands, face and eyes and consideration of clothing required. Pandolf et al. (1988) present detailed research and practical methods used for the assessment of outdoor environments by the US Army.

ISO 9920 (2007) provides a comprehensive data-base of clothing properties. ISO 13732-1 (2001) provides guidelines for when damage to the skin and burns will occur on contact with hot surfaces and ISO 13732-3 (2005) for damage caused by cold surfaces. ISO 13732-2 (2001) considers sensation caused by contact with moderate surfaces useful in the design of handrails, door handles, etc. All consider effects in terms of surface temperature, material type and exposure time (Parsons(2014, 2019).

Up mountains air pressure is lower than that at sea level, and special consideration should be given to the reduced oxygen content of air, increased evaporation rate, reduced heat loss by convection, cold and solar radiation (also reflected from snow). Clothing and equipment will also be important and for consideration by the environmental ergonomist. Parsons (2014) provides further information.

Adaptive opportunities for outdoor thermal environments include adjusting clothing and activity level, using sun-shades and more. Consideration and design of a space to include adaptive opportunities should use as a starting guide that human response to thermal environments is determined by the interaction of air temperature, radiant temperature, air velocity, humidity, activity level and clothing properties.

Sound

Outdoors sound is perceived by the auditory system and its measurement and assessment uses similar procedures, units and indices as described in chapter 8. For measurement outdoors, wind can affect microphones providing low frequency effects that can be reduced by using a foam 'fluffy' cover. Sound level decreases with the square of the distance from the source. Think of a point source power W (watts) radiating out as a sphere radius r (intensity $I = W/4\pi r^2$ Wm^{-2}) the inverse square law applies ($I \propto 1/r^2$). If it is interrupted by buildings, barriers, etc. then the sound is reflected, absorbed and transmitted depending upon the characteristics of the barrier.

The Day-night equivalent level (L_{dn}) for community noise and the Sound exposure level (SEL or L_{AE}) for single events such as trains or cars passing, are described in chapter 8. There are numerous guidelines and limits for outdoor community noise both generally and related to specific sources such as generators, construction work, music festivals and more.

The European Directive 2000/14/EC (see also Directive 2003/10/EC) considers noise emission by outdoor equipment and lists 57 types. Noise contours are often used to show noise levels over a community and around an airport for example. This may be the first stage of an outdoor environmental ergonomics noise survey where values of above around 55 – 65 dB(A) (although C, D (for aircraft noise) and linear weightings will also be useful) would provide a first indication.

Neighbour nuisance is complicated and works both ways (from hypersensitive neighbours to nuisance neighbours and all in between). Night-time (11.0 pm to 7.0 am) noise criteria are at lower levels (e.g.10 dB lower) than daytime limits (Haslegrave, 2015).

Vibration

Vibration outdoors is characterised in the same way as described in chapter 12 and is mostly about absolute perception thresholds (see Parsons and Griffin, 1988). Outdoor vibration is usually not expected and is almost always unwanted. It promotes dissatisfaction and fear of damage to property (often assessed by crack propagation in walls and ceilings) as well as alarm caused by earthquake tremors.

ISO 2631-2 (1989), is concerned with the evaluation of vibration in buildings, often transmitted from outside of the building, by trains, traffic, blasting for tunnels, construction work, etc. Criteria and limits are related to multiples of vibration levels above nominal absolute threshold values.

The operation of hand-held tools often takes place outdoors and it is recognised that effects will be more severe in cold conditions when the hands are cold. Vibration on rides at fun-fares provide special environments with rides to provide thrill and pleasure but not of damage to health. Low frequency vibrations (around 0.2 Hz) can cause motion sickness and is mostly related to vehicles. Further detail is given in Griffin (1990), and Mansfield, (2004).

Light

Outdoor daylight is from the sun and is at much greater levels than indoor light. It does not necessarily appear so, as the eye adapts to the level of light. Direct sun in the field of view causes glare as it is much brighter than its surroundings and looking directly at the sun will damage the eyes. Light levels outside are around 35000 lux in the summer and 8000 Lux in the winter, and indoors 500 lux is typical for an office.

Light is part of the electromagnetic radiation spectrum ranging from 380 nm to 760 nm. Radiation is the cause and light is the effect (Smith, 1995). It has both an electric and a magnetic field both varying sinusoidally with time. Infra-red radiation is at longer wavelengths than light and ultra-violet radiation is at shorter wavelengths than light. Both can damage the skin (ACGIH, 2024).

The cosine law for light states that if a light arrives on a vertical plane at 90° and the plane is rotated through an angle ø then the illuminance on the inclined plane is reduced by the ratio cos ø : 1. The inverse square law states that the illuminance E (lux) from a point source is proportional to the inverse of the distance squared $E \propto 1/(\text{distance})^2$. Light can also be reflected, absorbed and transmitted by objects.

Daylight is important as people prefer it to artificial light. An average daylight factor for rooms (%) may be of use to the environmental ergonomist; given by (T W Ø M)/(A(1-R^2)). Where T is the diffuse transmittance of glazing (0.85, clear single glazing; 0.75 double glazing); W is the net area of the glazing (m^2); Ø is the angle (degrees) subtended in the vertical plane by sky visible from the geometric centre of the window; M is a maintenance factor depending upon the cleanliness of luminaires,

A is the total surface area of the space (m^2) (floor, ceiling, walls and windows) and R is the average reflectance value of the internal surfaces (weighted by surface area).

So, for example, for the case study in chapter 18 (suppose a square room with 20 m walls × 4 m high) so A = 2 × 20 m × 20 m = 2 × 400 m^2 = 800 m^2 for ceiling and floor and 4 × 20 m × 4 m = 4 × 80 m^2 = 320 m^2 for walls and windows. so A = 1120 m^2. If T = 0.75; W = 40 m^2; Ø = 45°; M = 0.8; R = 0.5; so the average daylight factor is, (0.75 × 40 × 45 × 0.8)/1120 × 0.75) = 1440/1120 = 1.3 %.

Smith (1995) suggests that for average daylight factors of 5% and above, the interior will appear cheerfully lit by daylight. For values of 2% and below, artificial lighting will be required constantly. Although some subjective judgements and estimations are required the average daylight factor may provide supporting information for any environmental ergonomics lighting survey. It also demonstrates that good lighting design, as with good environmental design, is an art based upon science.

Boyce (2014) provides a comprehensive treatment of human factors in lighting. For driving he considers vehicle lighting, lighting in tunnels, road lighting, lighting for signs, signals and messages, and lighting in the rain, snow and fog. He also considers the use of lighting to reduce and prevent crime.

Air Quality

Outdoor air quality is mainly considered in terms of environmental health and pollution. It has been investigated throughout the world, particularly in major cities and often in relation to traffic (although coal fires in homes as well as industrial smoke have caused smog and much pollution).

Environmental health regulations have been implemented in many countries (European Commission, Directive, 2008/50/EC, air quality), (particularly with 'green; policies and the drive towards net zero carbon environments) and environmental air quality is continuously monitored with criteria and a variety of rating scales. As I write the river Seine in Paris is under investigation for pollution that will prevent an event at the Paris Olympics. It is not only air where pollution affects people.

Environmental ergonomists have generally not been involved with outdoor air quality. However subjective assessments using terms such as smelly, dissatisfaction and acceptability would allow air quality contours to be drawn in communities. Fanger (1988) suggests that the olf and the decipol could contribute to outdoor contours derived from the subjective percentage of dissatisfied values and an estimate of ventilation from the prevailing wind.

VEHICLES ENVIRONMENTS

Thermal

Vehicles provide microclimates that are greatly influenced by outside conditions. In particular the sun and weather. They differ from offices in that there are restricted adaptive opportunities, and that rapid changes can occur.

Hodder and parsons (2007) conducted laboratory and field research into thermal comfort in cars (see also Parsons, 2014). They proposed a modified index based upon the Predicted Mean Vote (PMV) index for passengers in vehicles exposed to direct sunlight. PMV solar = PMVshade + 200/RAD where RAD is the solar radiation from the sun ranging from 0, no sun, through 400 for cloudy to 1000 absolute maximum. Overhead sun in hot countries may cause discomfort but low direct sun for long days should not be underestimated in cooler countries away from the equator. The series of standards ISO 14405 parts 1 to 4 (2024) consider the ergonomics of the thermal environment in vehicles.

Small vehicles such as private cars require special attention. Larger passenger vehicles (trains, ships, aircraft) tend towards offices in consideration of thermal environments. Getting into and out of a vehicle provides a transient change in temperature for example, however the rapid change by moving out of or into moving air is of significance. Kelly (2011).

Passenger aircraft cabins are pressurised at around 8000 feet (2,440 m) above sea level, but this has little affect on criteria for thermal comfort (Nishi and Gagge, 1977, Parsons 2014). Restricted space, proximity of other passengers (Braun and Parsons, 1991; Parsons and Mahudin, 2004), the thermal properties of the seat and restricted adaptive opportunity, are all considerations.

Sound

Vehicles make sounds through their operation mainly through engine noise, exhaust noise, suspension response and also due to interaction with the environment including wind noise as well as wheel on road or track.

Noise emission from road vehicles is measured as a vehicle passes a sound-level meter under specified conditions. A value of 72 dB(A) for new cars in Europe is a typical upper limit, reducing to 68 dB(A) in the future and up to 85 dB(A) at 50 feet (15.24 m) in the USA.

It is unlikely that an environmental ergonomist would be involved in measurement, but for road noise, contour maps based upon noise measurement as well as subjective response would provide useful information. A similar assessment for aircraft can be made and regulations also exist. In the UK for example, a lowest observed adverse effect level (LOAEL) is 51 dB LAeq,16 h for an average summer day and 45 dB LAeq, 8 h for an average summers night.

Interior sounds in cars can be indicators of performance and provide feedback to the driver and include sounds from the engine and road as well as signals. In aircraft, warning and information signals are prevalent. Ship's engines are often more reassuring than disturbing and there are many internationally agreed maritime signals for correct operation of vessels at sea.

Vibration

Whole-body vibration is an integral part of vehicle environments and the methods and procedures in environmental ergonomics are described in Part 4. Griffin (1990) and Mansfield (2004) provide full accounts. Motion sickness is of particular concern especially at sea and in space and in road vehicles especially with children (and even simulator and virtual reality sickness where there is only apparent motion).

Sensory conflict theory is generally accepted as an explanation for motion sickness when one sensor, such as vision, is in conflict with another sensor such as the vestibular system. If the eyes indicate movement but the balance system says not then we are sick.

It is not surprising therefore that people in space are often susceptible to motion sickness (Reason, 1974). The environmental ergonomist could include subjective responses in motion sickness environments as well as motion sickness incidence. BS 6841 (1987) provides weighting function w_f with most sensitive vibration at 0.2 Hz, and Motion Sickness Dose Value MSDV = $(a_{wf}^2 t)^{1/2}$. Predicted % vomiting is one third of MSDV. Although applicable to sea journeys on ferries, the index may be applicable for other vehicle environments (Mansfield, 2004).

Light

Vehicle lighting provides a visual environment so that a driver can drive effectively and allows a vehicle to be seen. The environmental ergonomist can investigate both with appropriate measures. Guidelines and regulations are provided in Boyce (2014). Simple experiments can map out field of view.

In terms of vehicle lighting, the European Union requires that all luminaires of the same kind must be mounted with equal strength, equal height and symmetrical to the middle when mounted in pairs. European Union Directive 2008/89/EC is related to vehicle lights and vehicle signalling (EC, 2008b).

Air Quality

Vehicle air quality is usually related to exhaust emissions and is not usually the domain of the environmental ergonomist. Although smells and contour diagrams would be possible, it is usually chemical content and environmental health that are of concern. Cabon monoxide escape into vehicle interiors will affect performance and health. The assessment of air quality in passenger vehicles would benefit from environmental ergonomics surveys similar to those described in Part 6.

UNDERGROUND ENVIRONMENTS

Thermal

In an environmental ergonomics assessment of a coal mine I conducted some years ago, the weather was freezing on the surface, but on descent into the coal mine it was a constant 15 °C and did not seem to vary. Humidity was high due to watering to 'keep dust down' and tunnel surface temperatures were the same as air temperatures as the mine was not deep enough for significant geothermal heating. In some deep mines this is a significant cause of heat stress (Cooke et al., 1961). The geothermal gradient is that for every 40 m down from the earth's surface, there is roughly a 1 °C rise in temperature.

In the coal mine, air velocity was high to maintain ventilation. Although workers carried out strenuous physical work, I found no evidence of unacceptable thermal strain, and that the methods described in chapter 8 were appropriate for this application.

A few years later I was asked by a national safety executive to investigate heat stress in a tunnel that was being constructed for a new underground railway route. After descent on (staggered) ladders carrying equipment, we met the company safety officer. He insisted that as instructed he had increased ventilation levels. Using 'childrens' soap bubbles however we all watched as the bubble slowly drifted vertically to the floor. Air velocity was close to zero and ventilation was switched off or ineffective.

A follow up study used heat stress methods presented in chapter 8 and made recommendations accordingly. The work of the tunnel workers was using chipping hammers by hand to carve out the tunnel after the tunnelling machine had created the initial tunnel. In the case inspected the workers were digging out an underground station.

After the completion of the work I was asked to consider another part of the work where the tunnel ran next to the river (Thames). Compressed air was used in a sealed section of the tunnel and at a level that was greater than the pressure of the river at that depth to keep the water out.

It was necessary to modify methods for assessing hot environment to account for pressure. Workers were calibrated so that heart rate could be used to estimate metabolic rate and equations were modified in ISO 7933 (2004) to account for decreased capability to evaporate sweat and the increased heat loss by convection (Parsons, 1992; O'Brian et al., 1996).

Sound

Construction noise from underground working, as well as the operation of underground trains, can cause disturbance and the principles provided in chapter 8 will apply. Vehicles coming out of tunnels, particularly trains, will drive a compression wave that causes a boom when the train leaves the tunnel. Workers who construct tunnels will be subject to noise and hearing protection may be necessary (but may be uncomfortable due to heat).

The velocity of sound is related to the density of the air and in compressed air tunnels it will increase. However regulations for noise control will apply as will dB(A) and units with other weighting functions. Consideration should be given to infra-sound, low frequency noise and pressure waves.

Vibration

Vibration in mines and tunnels will be related to the operation of mobile machinery, blasting and holding of vibrating tools. Principles and practice for environmental ergonomics assessment will apply as presented in Part 4. Earth tremors and failure to signal blasting are likely to cause alarm due to fear of collapse.

Light

Enclosed mines and tunnels require light to ensure visual performance. Glare should be avoided and colour rendering is a consideration. Individual head lighting is used by miners and threat of fire and explosion are always present.

Tunnel lighting for traffic decreases from around 50 lux at the start of the tunnel to 30 to 40 lux at the centre of the tunnel. Facilitation of adaptation of the eyes is important. The bureau of mines information circular IC 9073, provides the underground coal mine lighting handbook (Rappaport and Szocik, 1986 and Lewis, 1986) and BS 6164 (2019) considers health and safety in tunnels.

Air Quality

Air quality underground is an essential consideration as oxygen is required to breath and toxic fumes must be removed. Effective ventilation systems are therefore essential. Systems of flexible plastic doors are often used to regulate air flow in coal mines. The environmental ergonomist can offer expertise in the assessment of smells but is not usually involved with the air quality aspects of health and safety where expertise is essential.

AQUATIC ENVIRONMENTS

Thermal

Cold water immersion can be exhilarating but can cause death on immediate entry or later through hypothermia. Avoidance is important and staying dry, as is being sensible, buoyant, hydrated, acclimatised, fat, determined, optimistic, insulated and sober.

Golden and Tipton (2002) provide guidance on survival in cold water. Water temperature and velocity (tidal stream) as well as sea state and weather, will be important and the environmental ergonomist may become involved in computer modelling, surveys or (ethical) empirical studies involving heart rate and internal body temperature and more. Predicted survival times are often requested and can be provided but depend upon individual characteristics and a range of parameters.

Diving is a specialist area involving protective, and maybe smart and heated, clothing, pressure, breathing apparatus and cold water temperatures. Physiological measures are appropriate in evaluation of effectiveness. Environmental Ergonomics studies of hyperbaric lifeboats (where divers can seek refuge underwater) show narrow temperature ranges for comfort with hypo- and hyperthermia occurring if there are small deviations.

Sound

Mechanisms of how sound will affect people with head above water are similar to those described in Part 3. Particularly with dry ears. Signals and communication will be affected by background (white) noise. Sound travels faster in water, and the sensitivity of divers to sound frequency and level will be different from that in air. Specialist investigation will be required for assessment and although environmental ergonomics may be useful it is not usually considered by environmental ergonomists.

Vibration

Vibration on the surface of water is mainly due to buffeting of waves and low frequency motion that can cause motion sickness. Under the water, operation of tools and machinery cause vibration directly into the body as mechanical vibration and a pressure wave will transmit in water (as sound). It is not known whether the same vibration criteria apply for whole-body and hand transmitted vibration underwater as for out of the water. An environmental ergonomist would be able to investigate using standard methods. A control group would probably be of use.

Light

Water in the eyes will affect vision, and solar radiation will provide daylight but also glare and reflections off the water surface. Sea state and low proximity to objects will affect the visual field.

Light will penetrate clear water down to 200 m but with a 50% loss in the first 10 m. The loss is due to scattering and starts with red light and ends with blue at the greatest depth. Pollution and particulates in the water (as well as lack of daylight) will reduce light levels and supplementary and personal lighting systems will be required for those operating underwater.

Water Quality

Water quality is not usually considered by the environmental ergonomist. Pollution caused by sewage, agricultural chemicals, ship discharge, plastics and much more could all be investigated in a survey but often require specialist knowledge, including medical expertise to investigate potential diseases.

IN THE AIR, IN SPACE, ON THE MOON AND BEYOND

Environmental ergonomics is the application of human response to the environment to the design of systems. By definition it is ubiquitous and in principle, it provides a wide scope that includes all environments.

Environments in aircraft require thermal comfort, in pressures ranging from sea level to 8000 feet (2,440 m, 75 kPa and 10.9 psi). Calculations suggest that thermal comfort indices used at sea level apply to aircraft cabins. Cockpits are bathed in solar radiation and sunlight and all aircraft are subject to accelerations and vibration, from the atmosphere and engines. A helicopter is particularly prone to noise and vibration. For sealed aircraft systems ventilation is important to maintain air quality, especially oxygen levels.

Space vehicles have similar criteria to aircraft environments and in principle applied ergonomics investigations are valid. A main consideration is a high level of acceleration on launch to escape the atmosphere and zero gravity when in orbit. Motion sickness is of major concern. Health effects are not apparent and do not appear to be chronic on return to earth.

The moon is becoming a major area for competition between nations, institutions and individuals. Speculation concerning human factors when living on the moon will turn into reality and environmental ergonomics has a role.

The moon has only one sixth of the gravity of that on earth. It has no atmosphere hence no weather, but it has extremes of temperature through radiation from the sun and then its absence. Space suits require carful design of the micro-climate to ensure survival. As there is no wind, etc. any buildings can be lightweight and in theory, made of silver tin foil with nothing to disturb them. That is unless meteors at the speed of bullets present a danger as they are not burned up in an atmosphere and account for the cratered appearance of the moon.

Much is known about how people respond to the physical environment and methods have been developed that are valid over a wide range of contexts. Diverse environments often require special consideration to adapt those methods to specific applications. A range of environments has been considered above, not to provide exhaustive information, but to provide 'starting' advice for the environmental ergonomist to consider how to approach an environmental ergonomics survey, if at all.

There is no formula for the design of an environmental ergonomics survey. It will be designed, using the methods available (subjective, behavioural, objective, modelling) to provide answers to the agreed, declared aims and objectives of the investigation. The following chapter recognises that not only are there diverse environments but that there are also diverse populations who occupy them.

FURTHER READING

Armstrong, L. E., 2000, *Performing in extreme environments.* Human Kinetics, Champaign, IL, USA. ISBN 0-88011-837-7.

Ashcroft, F. A., 2000, *Life at the extremes*, Harper Collins, London, UK, ISBN 0-00-255946-3

STANDARDS

See previous chapters.

22 Environmental Ergonomics in Diverse populations

INDIVIDUALS

There are over 8 billion people on earth and they are all unique. They also change with time. It is reasonable to conclude therefore that, at any point in time, they all have a unique human response to their environment. Although there are potentially 8 billion responses to any environment, the variation will depend upon the sensitivity of the measure we use. If we asked them all to say whether an environment is acceptable or not, and 5 billion say acceptable and 3 billion say not acceptable, the lack of variation is not due to lack of individual differences but lack of sensitivity of the measurement system, which is fine if a simple 'acceptable or not' is what we are interested in.

Although there are wide individual differences, as a species humans have a common genetic template. They also have varying experiences and cultures and many more attributes that allow them to be categorised into groups. Any categorisation will have inherent and deliberate bias and be related to the purpose of any investigation.

In environmental ergonomics, a global subject, if we select national geographical location as a category, it is relevant to know if thermal comfort criteria for people who live in China are the same as people from the USA, Europe and so on. Are dB(A); vibration dose value; visual sensitivity weighting curves; sensitivity to smells and more universally appropriate and are international standards truly international?

Chapter 1 outlined the dangers of assuming (without evidence) that, for example, people from hot climates require higher temperatures for thermal comfort (a convenient economic outcome!). Scientific evidence and consideration should not be replaced by commonly held beliefs and convenient assumptions and especially not on crusades to 'prove' a belief, however ethical, apparently humanitarian, or passionately held.

It is likely that the principles of determining human response to the environment, often based upon the laws of physics, are more universal than empirical studies with limited individual variation (e.g. principles such as, human response to vibration is frequency dependant, sweat required is a universal index, heat transfer will determine body temperature; A-weighted sound is related to annoyance; and so on). How much, but not the principle, may well depend upon individual and group factors. Methods may apply but environmental level at which an effect will occur, may depend upon groups.

It is important for the environmental ergonomist to recognise that experience, motivation, culture, social and political climate, psychological disposition and more

DOI: 10.1201/9781003401964-29

will influence the outcome of an investigation. The 'Hawthorne effect' is a well-known salutary lesson where people were so pleased that an interest was taken in their well-being, they reported greater satisfaction to all changes (e.g. lighting) because of the interest rather than their response to the environmental stimuli. The conclusion was that although lighting levels, etc. will have an effect, such things as productivity and satisfaction with the environment will be influenced by many other factors, including management style and worker relations, that influence outcomes, (Snow, 1927).

The four main methods of environmental ergonomics (subjective, behavioural, objective and modelling) including discourse analysis, focus groups and more should be selected with regard to the population under investigation. Subjective methods are sometimes not useful due to bias (disposition to please, inclination to determine an outcome, culturally not respected for opinions, fear of reprisal, etc.) or people not suited to use that method of communication (e.g young children, people with reading difficulties, etc.). Objective methods may then be used instead of, or as a complement to, subjective methods, that may use symbols rather than words and so on.

For environmental ergonomics it is important to consider individual differences in the population under investigation and that they may be multi-cultural, vary in gender or have illnesses and disabilities. Categories may have to be identified that have 'special requirements' or more correctly a range of different requirements. Disability discrimination legislation, for example, requires that working environments do not disadvantage people with disabilities.

A confusion occurred with the use of the term 'the handicapped' implying that some people are less capable in all things than others. Actually, no person is 'handicapped', however, all people can be handicapped by their environment and it is a challenge for the environmental ergonomist to provide environments where the population of interest (with a disposition to include as many people as possible, in inclusive design) can operate effectively.

Although all individuals are different, it is convenient to select categories to allow optimum environmental designs. When groups are selected, it is useful to identify their responses to the environment as a statistical population distribution, where parameters representing average and variation, provide useful practical information. Graphical representations are particularly important. It is also important to remember that parameters do not represent individuals, one of the ten fallacies in anthropometry is that 'there is no such person as an average person' (the fallacy being that there is). Pheasant and Haselgrave (2006).

ENVIRONMENTAL ERGONOMICS AND INCLUSIVE DESIGN

A full consideration of individual differences in Ergonomics, is provided in Wilson and Sharples (2015a) particularly in the chapter by Elton and Nicolle (2015) on 'inclusive design and design for special populations'. In the context of environmental ergonomics, inclusive design (see BS 7000-6, 2005) can be viewed as environments that are accessible to, and useable by, people with the widest range of abilities, within the widest range of situations, without the need for special adaptation or design. (e.g.

a large font on screens can be seen by all without adaptation for a person who is visually impaired).

Usability will ensure that environments can be used by specified users to achieve specified goals with effectiveness, efficiency and satisfaction in a specified context of use (see ISO 9241-210, 2010). Accessibility in environmental ergonomics refers to allowing users to access the features of environments through their sensory, physical and cognitive capabilities.

It is not the case that research into human response to environments has been restricted to fit, young, male, white, military, western people in laboratories, This may be a caricature, but it has been a typical starting population for environmental ergonomics research. Much has been researched into the diversity of populations and their requirements. But much is not known and an accepted philosophy of inclusive design is some way off Coleman (2001).

A challenge to the environmental ergonomist might be to design an environmental assessment to determine how people with atypical thermal responses (e.g. menopausal women), hearing impairment, atypical vibration sensitivity, visual impairment and olfactory impairment (e.g. loss in sense of smell) maintain health, comfort and performance. Elton and Nicolle (2015) provide a fuller discussion with practical application. Environmental ergonomics for specific populations is considered below in terms of national geographic location; age; gender; acclimatisation; people with disabilities and people with sensory impairment.

A method of providing optimum environments to individuals is to allow adjustment by individuals to the conditions to which they are exposed. In theory there will be no dissatisfaction (at least no complaint) if individuals set their environment to the conditions they want. In principle this should be successful but in practice it is not so easy. If individuals control the whole workspace then some will be satisfied and some not.

Early critics noted that one of the points of ergonomics was to avoid distractions due to environmental adjustment, ('fit the task to the person'). If the person has to fit themselves to the task then this would take time off task and reduce performance. Personal control systems have been produced for workstations but it is a technical and human factors problem to ensure individual environments do not influence adjacent workplaces and the environment overall. They may require training and cause even greater distraction than central control systems.

A more promising approach is to adopt the philosophy of designing for adaptive opportunity. Adaptive opportunities should be identified for the context, environmental components and their integration and the individual characteristics of the person.

Although there will be some distraction of people in the environment, building in adaptive opportunity allows individuals to create their own environments and is a worldwide challenge. It can ensure the health, comfort and performance of people across diverse populations and also provides an opportunity for low energy (green) solutions.

The philosophy is particularly important in consideration of diverse populations (e.g people with physical disabilities) and for all of the populations discussed below and more. As with the environmental ergonomics survey, there is no formula and innovation and inspiration, along with the general guidance and methods, are required. For people with special requirements it will be important to involve them in the design so that it meets their requirements.

National Geographic Location

It would seem reasonable to assume that people from different parts of the world with different climates, experiences, history, morphology and culture, will respond differently to (similar) physical environments.

Physiological adaptations such as in people living up mountains in hypo-baric environments, people used to sweating in the heat, behavioural adaptations such as knowing when and how to avoid environmental stress (based upon experience since childhood), people who are small with less protruding features (e.g. small noses and ears) allowing reduced heat loss in the cold and so on, all exist. There is no doubt that preferred tastes and odours vary across cultures and countries and that some physiological and perceptual differences apply. The concept of smell however seems to apply even if different odours produce different responses across populations.

The important point is whether differences are sufficiently significant in terms of environmental ergonomics and the design of environments and whether it is reasonable to assume responses are from a homogeneous population.

It maybe that populations physiologically, morphologically and behaviourally adapted to environments can tolerate more environmental stress than those who are not. Even that is for debate as people in hot climates avoid hot times of the day for work, people in the cold avoid being cold. Vibrating tools show increased damage to hands not because of individual 'national and regional' differences but because environments are cold, people can be acclimatised to heat just as effectively if they are from moderate or hot climates (Clark and Edholm, 1985).

There is little evidence that thermal comfort or performance requirements for environments differs between those from different national-geographical populations. (e.g. Fanger, 1970) or that hearing damage or frequency weightings differ across populations and more. It is also clear that more research is needed.

International standards have been developed by international experts and produced and accepted by international vote, to apply world-wide. Principles concerning human response to the environment are usually adopted globally but regulations, guidelines and limits sometimes vary.

It has been my experience that despite many attempts to identify specific requirements and exceptions from international standards, there has been a surprising commonality across populations in terms of human response. A practical starting point for the environmental ergonomist concerned with international populations is to begin by assuming international standards apply and investigate and evaluate evidence that they don't, if it is suggested. The environmental ergonomics survey will provide some information and it is often possible, as part of the survey, to test whether there are significant differences in the responses of groups and individuals.

Age

As physiological and cognitive faculties age, they deteriorate. That is, to different extents in different people and at different times. It is important for the environmental ergonomist to identify, and take account of, the age of the occupants of the environment under consideration.

Human perception may be dulled by age, although experience can provide more effective behavioural responses (e.g. motor car and industrial accidents are more prevalent among the young). Tolerance to environmental stress reduces with age even though fitness can compensate and is important (Havenith and Van Middendorp, 1990).

On most continents birth rates are falling and in Africa they continue to rise. Due to migration national populations are often rising and provide an influx of young people and multi-cultural societies and a challenge for the environmental ergonomist.

The statement that the reason populations have increasing percentages of elderly people is due to increased health care has ceased to be valid. Average age of death remains higher for females than males, neither are rising (Raleigh, 2024).

Falling birth rates may be a complex relationship between poor maternity care and fear of giving birth, the desirability and necessity of having children, social norms and fashion, emancipation of gender, loss of 'freedom', as well as the usual argument that standard of living is reduced by the expense of children, and more.

A general consequence of age is vulnerability to environmental stress due to deterioration, often causing strain on the heart. For thermal conditions, reduced thermoregulatory capacity, and often denial by the individual leading to unsafe behaviour (especially in the heat), can lead to fatal exposures.

Probably because of reduced physiological requirements, conditions for comfort are not affected by age, although as Fanger (1970) points out, the elderly are less active with lower metabolic rates, leading to higher temperatures for comfort. But in comfort assessment, metabolic heat production is an independent variable so included in any evaluation and should be taken into account in any estimation by the environmental ergonomist.

The amount of light falling on the retina is greatly reduced in the elderly and deterioration of the eye and visual system, including visual acuity and spectral sensitivity, is a common effect of age where spectacles and maybe contact lens' are common and almost always needed for reading in later life.

Boyce (2014) considers lighting for the elderly. He notes that as well as less light reaching the eye and diminished ability to focus close-up, scattering and stray light inside the eye is increased and spectral qualities reaching the retina change. All of these effects, often increasing from early adulthood to old age.

These lead to reduced absolute sensitivity to light, reduced visual acuity; contrast sensitivity; colour discrimination, and light adaptation. So reading in dim light, reading small print and distinguishing between colour all become difficult. As Boyce (2014) points out, 'these are the best we can expect' and do not include common pathological changes (see below). The environmental ergonomist must consider changes with age and maybe it will be a case of 'fitting the person to the task' and screening for people who may be at risk.

The effects of age on responses to sound include impairment of hearing level and frequency. This can be caused by deterioration of the auditory system, including the inner ear and auditory nerve. Most notable effects are a reduction in the ability to hear high frequencies and speech, and distraction due to background noise and sometimes tinnitus.

Sensitivity to vibration at the skin decreases with age particularly at high frequencies (e.g. 250 Hz). There is no consistent effect of age on responses to whole-body vibration and often suggestions that it can improve muscle strength and balance. Care must be taken however as people with chronic illnesses and disabilities may be harmed by whole-body vibration.

Sensitivity to taste and smell deteriorates with age. The ability to taste sweet and salty foods deteriorates first. The sense of smell also deteriorates with age and illnesses such as COVID (coronavirus) can damage the sense of smell and taste.

Gender

Research into the effects of gender on human response to the environment has divided the population into males and females. A pragmatic starting position is that there are no gender differences in response, that international standards apply and that the environmental ergonomist should consider gender issues explicitly in the context of any investigation. In addition factors such as type of clothing and anthropometry will also be considerations.

Conditions for thermal comfort have generally been found to be similar for males and females (Fanger, 1970) although deviations from comfort conditions, especially towards cooler environments, have shown that (for identical clothing and sitting at rest) females become unacceptably cold at higher air temperatures (etc.) than males (Parsons, 2014).

Clothing insulation and design are important, especially when considering draughts. Females often have thinner fingers than males and this may influence responses to cold (Parsons, 2014). They are also more inclined than males, to report dissatisfaction when conditions vary from neutrality (Karjalainen, 2012).

It is commonly held that menopausal women have varied thermal responses and despite its importance there seems to have been little research into optimum thermal conditions for menopausal women. It is logical that pregnant women in the later stage of pregnancy will have increased metabolic rates during activity, reduced adaptive opportunities and may be disrupted by hormonal changes.

Thermal tolerance may show differences across populations but these may reflect cultural differences. I found that Chinese women seemed to be more tolerant to heat in a thermal chamber than males and western females (see also Havenith et al., 2020). That is not to say that they would be dissatisfied with international thermal comfort standards.

Females may be more sensitive to cold stress than males, probably because they have a greater surface area to mass ratios, and smaller fingers despite having greater proportions of sub-cutaneous fat than males. Studies of heat stress have found no differences between males and females if acclimatisation and fitness are taken into account (Havenith and Van Middendorp, 1990).

There appear to have been few studies on the effects of gender on responses to noise and vibration that can provide general conclusions. Logical considerations do not suggest gender differences for hearing and that annoyance will be related to psychological disposition and social context. Body shape and size may affect responses to both hand transmitted and whole-body vibration, but although there have been studies (e.g. Griffin et al., 1982; Parsons and Griffin, 1983), no systematic effects of gender have been identified.

Boyce (2014) considers the way ahead for lighting research and notes that

> ...to achieve a more fundamental understanding of how people respond to lighting, it is necessary to put people first and the lighting conditions second... The people considered have to be representative of the actual users in all their variety and complexity.
>
> *Boyce (2014)*

He also emphasises the importance of context. It can be concluded therefore that, as with other environmental components, the effects of individual differences, including gender, require further research.

Gender differences and the effects of smell have been identified in laboratory studies often related to sexual attraction. Females have been found to be more sensitive to smells than males and report smells more frequently. The menstrual cycle affects sensitivity to smells but for the environmental ergonomist gender differences in sensitivity are not of sufficient practical significance to be of primary consideration.

Cultural, social and behavioural responses as well as dissatisfaction however may be of practical importance and should be considered within the context of the investigation. Napoleon's message to Josaphine 'home tomorrow don't wash' is not a universal preference!

Acclimatisation

Repeated exposure to heat, or long term exposure to hot climates trains the thermoregulatory system to become more efficient, particularly in the increased production of sweat. Acclimatised people do not require different conditions for thermal comfort (Parsons, 2014) and cold acclimatisation is debatable and should not be assumed. Behavioural acclimatisation (instinctively or consciously knowing how to behave appropriately to reduce thermal strain) is of particular importance. For example, the adjustment of clothing or the avoidance or anticipation of danger.

Loud sounds cause a temporary threshold shift that alters the impedance of sound transmission in the ear to provide protection (Elgstrand and Petersson, 2009). The increased ability to filter out background noise and unwanted sounds is also a form of adaptation.

Vibration can also cause a temporary threshold shift in hearing and also a shift in sensory perception. (i.e. a reduction in vibrotactile sensitivity and nerve conduction velocity).

Responses to light by the eye have a built-in adaptation as the amount of light reaching the eye continuously adapts to the average level of light. When moving from dark to light conditions there is rapid light adaptation and from light to dark, a slower dark adaptation. There seems to be no evidence that training can improve the ability to adapt to changes in light level.

Acclimatisation may universally occur with increased and frequent exposure to an environment (e.g. in training) and physiological response (e.g. more rapid threshold shifts in physiological adaptation) and a fuller understanding is required. Acclimatisation, complements the human disposition to optimise responses to the environment and knowledge of effects may emerge in the future.

People with Disabilities

There are no general guidelines on optimum environments for people with mental and cognitive impairment. A first stage is to assume that they are similar to those for people without impairment. Particular attention should be paid to adaptive opportunities and ability to take advantage of them.

Parsons (2021) provides a summary of research into thermal comfort conditions for people with physical disabilities. People with cerebral palsy; spinal injury; spinal degeneration; spina bifida; Hemiplegia; stroke; Polio; osteoarthritis; rheumatoid arthritis; head injury; and multiple sclerosis were exposed to warm; neutral and cool conditions and in general it was found that conditions for thermal comfort were similar to those for people without physical disabilities.

People who are deaf or with hearing impairment or blind with visual impairment will require special consideration. It seems logical that vibration will affect people with disabilities more than those without disabilities but there have been no studies. There is no evidence that people with disabilities will require different olfactory environments from those without disabilities and caution must be taken if it is assumed that people without the sense of smell can be exposed to otherwise smelly environments.

People with Sensory Illness and Impairment

Specific illnesses related to responses to environmental components are not usually the concern of the environmental ergonomist and will often be considered when screening for people fit for work. Medicinal and recreational drugs, including alcohol, will have effects on human response to the environment and its reporting. It is prudent to seek medical advice.

For completeness examples of effects are summarised below. People with illnesses occupy environments and may become the concern of the environmental ergonomist.

Thermal: Thermoregulatory failure: drugs used in medication (including blood thinners and blood pressure drugs) can reduce ability to sweat and affect blood flow, and in environmental extremes the environmental ergonomist must be made aware of use among those exposed to heat or cold. Diseases can also affect thermoregulation as well as infections and fever. A list of contra-indicators is shown in Appendix 2.

Sound: Hearing loss can be temporary and permanent and can be caused by exposure to loud noise as well disease, genetic effects and increased age.

Vibration: Raynaud's phenomenon is a disease that causes reduced blood flow in the hands and feet. It has genetic connections and is exacerbated by cold environments.

Light: Low vision is a state between normal vision and total blindness and is caused by cataract (opacity of the lens); macular degeneration (opacity in macular in front of the fovea); glaucoma (damage to blood vessels supplying the retina, due to pressure, causing narrowing of the visual field); and diabetic retinopathy (destruction of part of the retina).

Air quality: A consequence of having COVID is a loss or reduced ability of being able to taste and smell that may become long-term.

The above descriptions are examples of how people in a populations of interest to the environmental ergonomics survey may be diverse and require special considerations. A full environmental survey will assess the integrated environments and will consider interactions of environmental components as well how individual components add up to the whole. A full description with examples is provided in Part 8.

FURTHER READING

Wilson, J. R., Sharples, S., 2015, *Evaluation of human work* (4th Ed), CRC Press, Taylor & Francis Group, Boca Ratan, New York, Oxford, ISBN 978-1-4665-5961-5

STANDARDS

ISO 28803, 2012, *Ergonomics of the physical environment - Application of international standards to physical environments for people with special requirements*, ISO, Geneva

Part VIII

Environmental Ergonomics and Integrated Environments

Part 8 presents the environmental ergonomics assessment of total environments in two chapters. Chapter 23 presents principles, methods and ideas for the assessment of total environments and 'brings the subject together'. Chapter 24 describes an assessment of an integrated environment in an office that has been designed to represent the highest standard possible.

Chapter 1 introduced the subject of environmental ergonomics, and defined it in the context of Ergonomics, as the application of knowledge of human response to the environment to the design of systems. Chapters 2 and 3 introduced environmental ergonomics methods. Specialists in components of environments such as noise or light will have been comfortable with specific subsequent chapters and hopefully inspired by others, but it is the whole in the context of a system, and not the component parts that gives environmental ergonomics its unique contribution. Part 8 returns to the role of the environmental ergonomist in assessing total, integrated environments. It covers all of the environmental components and how they are brought together.

The possibility of an environmental index that represents the quality of a total environment is discussed and is a challenge for the future. Just as environmental ergonomics should draw the line at perception of air quality and not include the effects of 500 or more chemicals and biological exposures; so we should decide on the extent to which pleasure, aesthetics, inspired environmental design and so on fall within scope. In principle yes, in practice it requires specialisation.

DOI: 10.1201/9781003401964-30

We can speculate that a 'total' environmental index will require a common evaluative construct across all environmental components to give the index value. That would provide a good starting point for combining the effects of different environmental components. A model of the effects of the total environment would be a significant challenge, however, predicted dissatisfaction is common for individual components and if we add extremes of intolerable and delight we may have a workable scale (Figure P8). Computer models for each component may take the lead from thermal comfort and attempt to predict the mean vote of a large group of people on the scale provided. Alternatively, measured values my take the mean of the ratings on the scale, provided by those exposed to the environmental component.

It would be valid to use such a scale and ask individuals to rate how they find the whole environment. An investigation into why that rating was given could then follow. The concept of an index is to provide a single number however and using the same rating scale for thermal; sound; vibration; light; and air quality and a method of integration to combine the numbers to a single environmental index value will also add information of why we arrive at that value and provide focus on methods for improvement.

Allowing influence of all environmental components on the final index value feels prudent. Taking the root of the sum of squares of component values, would provide more weight to extremes but allow influence for all. The root-mean-square (rms) value would provide a value related to the scale. However, intuition suggests that it is the worst-case value over all components that would determine overall environmental quality especially if one of the components is deemed intolerable.

Such an index could be tested subjectively by comparing the individual component method with overall ratings in a controlled study. If future research leads to predictive methods for each component (predicted satisfied or dissatisfied) then the link between the measurement of the physical environment and the total environmental index will be made for evaluation and assessment as well as computer aided

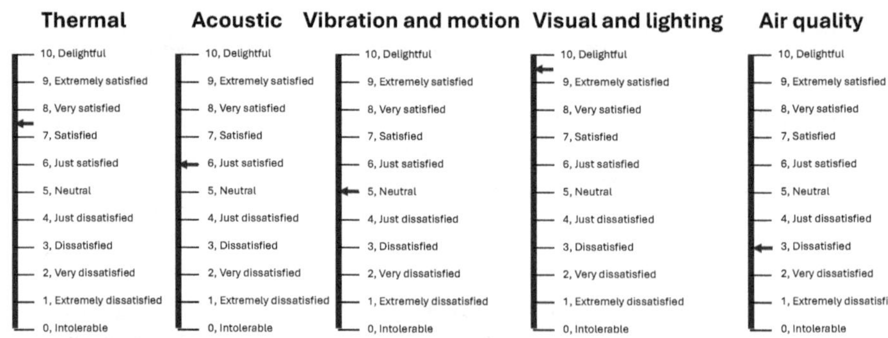

7.5 6.0 5.0 9.5 3.0

Environmental index (EI): rms method = $((7.5^2 + 6.0^2 + 5.0^2 + 9.5^2 + 3.0^2)/5)^{1/2}$ = 6.6 (satisfied)
Worst component method = 3.0 (dissatisfied)

Minimum threshold:
Gold , EI > 8.0 (very satisfied) : **Silver**, EI > 6.5 (just satisfied to satisfied) : **Bronze,** EI > 5.0 (Neutral)

FIGURE P8 Environmental index (EI) as an indication of the quality of the environment. (EI > 9.0, Platinum; delightful?).

design, with manipulation of variables, scenario testing, standardisation, labelling (gold-standard!) and more.

Chapter 23 considers integrated (total) environments and builds upon previous chapters concerned with individual environmental components. People are surrounded by total environments that can be represented as the integration of environmental components and their interactions. Consideration is provided for the development of a total environmental index. It could be used to provide a quality mark for environments. As for environmental assessment of environmental components, methods for assessing total environments relate to subjective, objective, behavioural and modelling. Assessments are made by component and brought together for the integration. The Hawthorne effect is noted, where people were positive about any change to their environment because someone was taking an interest in them. Methods of combining the effects of environmental components and their interaction are presented. Cross modality matching as a technique is described and homeostasis in terms of the total environment is discussed. ISO 28002 (2012) is the first standard to consider the environmental ergonomics survey of total environments. Proposals for measurements of the environment and subjective scales for measuring human response are described.

Chapter 24 provides a challenge for the Environmental Ergonomist as the requirement was for the best that can be achieved with no expense spared. Joshua had been given the task of attracting the top people in the world to work in an office on the west coast, The project was of top priority and it was imperative that the workers were happy, felt valued and were productive. He contacted Wally, an Environmental Ergonomics professor who had set up a 'spin off' company with his successful PhD students. After some discussion of terms of disclosure agreements a contract was signed. The aim was to investigate whether the office as set up could be regarded as 'gold-standard' and if not, to make recommendations to ensure that it was. Joshua needed to report to his senior managers that this was the case.

The survey was mainly conducted remotely as the investigators were not allowed direct contact with the workers. Systems were tested in the environmental ergonomics laboratory and a pilot trial was conducted to ensure that all required data would be obtained and meet the requirements of the project objectives. Trolley systems containing instruments were positioned across the room to measure the environment at workplaces. Existing overhead cameras allowed observation measures. Thermal, acoustic and vision and lighting measurements were taken on the trolley and accessed by the internet for analysis. Vibration was measured with accelerometers on the floor. Air quality was assessed using a panel of female judges who entered the room towards the end of the day and judged the air quality. Subjective measures were obtained via a computer interruption questionnaire to each worker at specified times.

The overall finding was that the office environment was of high quality and that workers were very satisfied. The thermal environment could be regarded as gold-standard. Computer noise and vibration were acceptable but too high to be regarded as gold standard. There were some reflections in the visual field and the décor obstructed the view. Light levels could however be regarded as excellent. On receipt of the report it was interpreted as confirming gold-standard and reported to senior management. Actions were taken however to remove the computer processors, but not screens, from the office and picture frames were modified to avoid reflections.

23 Total and Integrated Environments

TOTAL ENVIRONMENTS

It is a truism that people exist in external environments as they are defined as the environment surrounding the body. The effective design of systems involving people therefore must take into account the response of the people to the environments in which they operate. They are always present.

The environments are total environments and it is the integrated effect of the total environment that will influence human response. For convenience the total environment is considered by the ergonomist as an integration of environmental components (thermal, noise, vibration, light, air quality, etc.). It is also often the case that one component is of particular concern (e.g. thermal discomfort) and becomes the starting point and focus of any environmental ergonomics investigation.

For a comprehensive assessment of a total environment all components may be considered, to provide an overall measure of the environment. A total environmental index (e.g. integrating all relevant variables into a single number relating to the quality of the environment) would provide a quality mark or more generally be related to the total effect. Such an index has been proposed but not developed to date and was considered as early as the 1960s. Environmental assessments of total environments will provide an overall initial assessment of an environment that will allow a focus on particular concerns and components for future investigation.

The environmental ergonomics survey will concentrate on the assessment of how the total, integrated, environment will influence the health (and safety), comfort (including inspiration and pleasure) and performance (physical and cognitive) of people who occupy the environment and will involve physiological, psychological, social and behavioural responses.

As always, we must remember that any investigation may interfere with what it is trying to measure (e.g. The Hawthorne effect – Snow, 1927) and may raise concern, provide reassurance, demonstrate care, raise individual and group esteem and group cohesion, provide a caring social interaction (others, such as investigators and managers, care about what you think and feel) and more, all of which must be judged in the context of the investigation.

ENVIRONMENTAL ERGONOMICS METHODS AND THE INTEGRATED ENVIRONMENT

Environmental ergonomics methods for measuring human response, can be considered in terms of subjective, objective, behavioural, social and modelling methods. Physical measures of the environment must use appropriate instrumentation to

DOI: 10.1201/9781003401964-31

quantify the variables important to each environmental component. There are collections of tools (toolsets) that achieve this but no instrument that integrates the measures to provide a total environmental measure, although energy could be considered to provide the basis for a common unit.

Behavioural measures in response to environmental discomfort will depend upon the nature of the discomfort. Adjusting clothing insulation when too hot or too cold is a common response to thermal discomfort. Behavioural responses will be observed and interpreted in the context of the study. Moving away from a space would seem to be a common behavioural response to all individual environmental components and to an integrated environment. Common subjective responses to individual component and total environments include comfort, satisfaction, pleasure and so on and heart rate is an objective measure that could be regarded as an integrated response to all environmental stressors.

Social cohesion may be a measure of group response to integrated environments (e.g. sociograms in classrooms). Computer and physical models of human response to the environment are concerned with individual environmental components. Although attempts have been made to combine knowledge in computer aided environmental design software (Smith and Parsons, 1987; Wadsworth and Parsons, 1989).

COMBINED ENVIRONMENTAL COMPONENTS

If two or more components of an integrated environment significantly affect human response then it is part of the assessment to consider how the effects will combine to give a total effect. Synergy suggests that the combination may provide an effect that is greater than the sum of the parts, masking may provide a model where the dominant component with a contribution from others will provide the response and logarithmic, exponential and power summations (e.g. addition of the fourth power and more of the (quantifiable) effect provides a dominant contribution of the peak component) may be valid. A pragmatic model is to assume that the most severe component dominates human response irrespective of responses to other components.

Much will depend upon which particular effect is considered, for example health, comfort, performance, distraction, satisfaction, pleasure and so on. A related point is that it may be the same person that is dissatisfied by all environmental components. So if there are 100 occupants only 1% will be dissatisfied with the environment not 5% for 5 components and so on. This point was made for combining dissatisfaction due to local thermal discomfort (e.g. draughts) and overall thermal discomfort where it was stated that percentage dissatisfied predictions for each individual affect, should not be added (ISO 7730, 2024).

INTERACTION OF ENVIRONMENTAL COMPONENTS

The effects of combined environmental components is different from the interaction of environmental components. For example if Light causes effect L and Noise causes effect N then the effect of both together will be L + N + L × N. Where L × N is termed the interaction of the environmental components and is an effect uniquely caused by the presence of both light and noise together. If the effects of L and N combine in a consistent way, so that the effect of L does not depend upon the level of N and vice versa, across all levels of L and N then there is no independent interaction. If we develop the model using a simple statistical linear model approach, and add in vibration with effect V, then we have the overall effect $E = L + N + V + L \times N + L \times V + N \times V + L \times N \times V$. For five environmental components making up an integrated environment there will be 5 main effects and combinations up to a five-way interaction and so on.

An example of an interaction would be that for a given level of odour, the dissatisfaction caused will depend upon the air temperature as, for air containing the same constituents, cool air is perceived as being fresher than warm air. Fanger (1970) reviewed studies into the interaction of colour with thermal comfort requirements. People perceive blue as a cool colour and red as a warm colour. However when tested in a thermal chamber the colour of the walls had no effect upon the physical conditions required for thermal comfort. Painting the walls blue will elicit cool emotions but will not save on energy used for air conditioning. I found a similar effect when I played (large screen) movies containing cold people in the snow or movies with people in the hot desert to my students in a thermal chamber. There was no difference in conditions they judged as thermal comfort, cool or cold.

After much consideration world-wide and some experimental studies (e.g. Grether et al., 1971), it can be concluded that for most practical purposes, interactions of environmental components can be ignored or are so few that they require special consideration (such as the case of cold discomfort and air quality). A note should be made of interaction in terms of response. Hot conditions produce sweating which increases the bio-effluents (smells) produced by people. This will have consequences for air quality but is not necessarily an interaction of the effects of environmental components. It increases the level of the main effects of one of them.

As there are a limited number of interactions, that can be considered as special cases, it is reasonable for most applications to consider the most severe environmental component as the area for recommendation for improvement and further investigation if required.

If all contributions are included and the effect of each component can be quantified (e.g. discomfort rating) then a power addition may be appropriate. Huang and Griffin (2014) suggested a root-mean-square addition for combining noise and vibration. For all components maybe a fourth power method would be more appropriate allowing dominant components to have dominant weightings. This could form the basis of an environmental index with further evaluation needed as well as consideration of how to calculate the effects of exposure time and a 'dose' value.

CROSS MODALITY MATCHING

Cross modality matching is where the effect of one (or more) stimulus in a particular environmental component (mode) is compared (subjectively) with the level of another stimulus in a different mode, so that the relationship between the two can be determined.

For subjective assessment, in the assessment of total environments, each individual component could be rated in terms of discomfort (or other) relative to a fixed stimulus in a selected mode, for example 60 dB(A) white noise, to give a common scale for combining effects. Richards et al. (1978) used this technique to investigate the ride quality of buses and trains.

Laboratory studies have investigated combined effects with cross modality matching (mainly noise and vibration) but for application in environmental ergonomics the subject is in its infancy. Huang and Griffin (2012, 2014) present results where noise and vibration were judged individually and in combination to provide a root-mean-square relationship and Howarth and Griffin (1991) investigated the combination of noise and vibration recorded in houses near railway lines from trains and showed how the vibration dose value and the sound exposure level could be combined to provide a prediction of the overall effect on annoyance. Flemming and Griffin (1975) derived curves of the relationship between noise and vibration in terms of when presented in combination, which would participants prefer to have reduced.

ASSUMPTIONS ABOUT THE PEOPLE RESPONDING TO THE ENVIRONMENT

Environmental ergonomics is primarily a practical subject that uses knowledge of human response to the environment to design (including evaluation and assessment) environments as part of systems. It is important however to take a fundamental view of the human disposition, in terms of any individual persons' response to any environment, as this will provide a perspective on the people involved in the system, the primary user-centred consideration of environmental ergonomics.

The subject is complex and a full understanding is not known, however a useful underpinning model of a person in an environment provides a starting point sufficient for effective environmental ergonomics investigations.

Parsons (2014) made this point in his book 'human thermal environments' and it can be paraphrased and extended for environmental ergonomics to note that people are not simply a body mass controlled by physiological systems that mechanically respond to environmental stimuli. They exist in a social context and have aspirations, ambitions, emotions, worries, opinions and concerns and consider the past, present and future. They interact and communicate with others, instantly near and far, verbally, visually and by electronic systems (mobile telephones, 'tablets' and computers). They have an image of their 'self' and how other people view them. They are active, learn from experience, and have expectations, memories, predictions and models of the world. They test scenarios and predict the future, and have religions, beliefs and cultures as well as social networks, and they engage in discourse in their interaction with others and a global society, not to mention bias, prejudice, political motivation, compatibility and incompatibility with others including competitiveness, as well as family and friendship bonds and more.

As Alexander Pope put it in his essay on man of 1734, "A formidable task but we have a plan". A comprehensive study of human environments, including physiological, psychological and social considerations, must consider people as a whole.

Wilson and Sharples (2015a) suggest that the combination of methods used in ergonomics investigation will depend upon context and imply that there is no 'recipe to follow' and that the skill and experience of the ergonomist will be important. Parsons (2014) in his discussion of thermal models concludes that the model used (adopted for the investigation) should be the one most appropriate for the investigation under consideration. He notes that no model will be perfect and that on the selection of which model to use, "The question of whether a model 'works' or not becomes a question of whether the imperfections are significant in terms of the application to which the model is put.

ENVIRONMENTAL ERGONOMICS AND HOMEOSTASIS

Ramsay and woods (2014) provide a review of homeostasis and its conceptual development in physiology. Claude Bernard (1870) noted that the body maintains an optimum internal environment independently of the external environmental conditions and perturbations. He makes a general point but we should note here that the internal environment includes temperature, blood sugar, pressure, volume, osmolality and glucose, body fat and a multitude of other regulatory systems and their interactions. In general terms, we could also include the amount of light entering the eye, reduction of auditory transmission in the ear and so on. The body therefore regulates its internal environment.

A general observation can be derived from an extension of the work of Cabanac (1981) where thermal pleasure is experienced when cooling down if too hot or when heating up if too cold. The general principle seems to be that if a person moves towards and experiences conditions that are perceived as desirable (e.g. homeostasis) then the body is rewarded with a feeling of pleasure.

Pavlov (1849–1936) and his dogs added to this by demonstrating that the body also regulates its internal environment in anticipation of an environmental change. The conditioned reflex in anticipation of food can be extended to the (subconscious) use of experience to provide an optimum internal environment not only in response to stimuli but in anticipation of stimuli. Canon (1929) was first to propose the term 'homeostasis' and included the 'wisdom of the body' not only to respond to stimuli but to anticipate them.

All can be considered autonomic and sub-conscious actions. The internal environment can therefore paradoxically be regarded as a set of continuously changing variables providing an interacting environment within optimum ranges of the variables depending upon requirements, all without the conscious intervention of the person.

Parsons (2020) extends homeostasis by, in addition to subconscious physiological responses, adding in conscious, behavioural responses so that a person can occupy a perceived optimum environment. For example, perceived discomfort, safety, or to avoid loss in performance. This involves responses to both anticipated and actual stimuli.

Parsons (2020) suggests that in the context of thermal environments, perceived thermal comfort can be regarded as the regulated variable, with physiological regulation providing autonomic responses ensuring survival. For environmental ergonomics in general, it is reasonable to hypothesise that the regulated variable could be assumed to be perceived and anticipated discomfort to preserve comfort.

ISO 28002, 2012, ERGONOMICS OF THE PHYSICAL ENVIRONMENT – THE ASSESSMENT OF ENVIRONMENTS BY MEANS OF AN ENVIRONMENTAL SURVEY INVOLVING PHYSICAL MEASUREMENTS OF THE ENVIRONMENT AND SUBJECTIVE RESPONSES OF PEOPLE

In recognition that an environmental ergonomics survey is an internationally recognised tool for ergonomics investigation and design, ISO 28802 (2012) was developed to provide guidance that will allow the starting point for the design and practical application of a survey in the context of the environment under consideration.

> *This International Standard provides an environmental survey method for the assessment of the comfort and well-being of occupants of indoor and outdoor environments. It is not restricted to any particular environment, but provides the general principles that allow assessment and evaluation.*

Parsons (2015a) provides a description and a case study of an environmental ergonomics survey of an office where people refused to work and Chapter 24 presents a case study involving a post occupancy assessment of offices in a new organisation.

Table 23.1 summarises the physical and subjective measures suggested for the assessment of single environmental components to total environments for thermal;

TABLE 23.1
Physical and subjective measurements (adopted from ISO 28002, 2012)

Environmental	Measure	Standard	Subjective term	Comment
Air quality	CO_2 %dissatisfied	instrument specification	Smelly	%dissatisfied for decipol and ventillation rate for olf.
Vibration	awrms VDV, A(8)	ISO 2631 ISO 8041	Discomfort	Integrate values across axes and Inputs. Weighting function for application: motion sickness, discomfort, hand etc.
Vision and lighting	Horizontal illuminance lux	CIE 69	Bright-Dark Discomfort	General impression glare, flicker, reflections interesting or not…
Acoustic And noise	dB(A) dB(A)Leq	IEC 61672-1 ISO 9612	Annoying	85 dB(A)Leq and 137 dB(C) peak circa damage limits. Circa <55dB(A) for offices.
Thermal	ta; tr; rh; v	ISO 7726		+3,hot, +2,warm,
	Clothing	ISO 9920		+1,slightly warm, 0,neutral
	Activity	ISO 8996		-1,slightly cool,-2,cool,
	Comfort	ISO 7730	Sensation	-3,cold
	Heat stress	ISO 7933	Core temp, sweat	physiological measures
	Cold stress	ISO 11079	Core and skin temp.	ISO 9886.

acoustical; vision and lighting; air quality and vibration components. The standard relates to the comfort and well-being of people. It should be noted that environmental ergonomics investigations will also include effects on health and performance.

For each environmental component physical, subjective and observational measures are presented and an annex to the standard provides an example of an environmental ergonomics survey in a building. Analytical methods are presented by reference to relevant standards and a bibliography provides references related to the assessment of individual environmental components and 'total' integrated environments (Table 23.2 and Appendices 3–13).

TABLE 23.2
References providing information for the integrated environmental ergonomics survey (ISO 28802, 2012)

1. ISO 1996-1, Acoustics — Description, measurement and assessment of environmental noise — Part 1:Basic quantities and assessment procedures
2. ISO 1996-2, Acoustics — Description, measurement and assessment of environmental noise — Part 2: Determination of environmental noise levels
3. ISO 1999, Acoustics — Estimation of noise-induced hearing loss
4. ISO 2631-1, Mechanical vibration and shock — Evaluation of human exposure to whole-body vibration — Part 1: General requirements
5. ISO 6385, Ergonomic principles in the design of work systems
6. ISO 8995-1, Lighting of work places — Part 1: Indoor
7. ISO 9241-6, Ergonomic requirements for office work with visual display terminals (VDTs) — Part 6: Guidance on the work environment
8. ISO 9241-7, Ergonomic requirements for office work with visual display terminals (VDTs) — Part 7: Requirements for display with reflections1)
9. ISO 9241-8, Ergonomic requirements for office work with visual display terminals (VDTs) — Part 8: Requirements for displayed colours
10. ISO 10551, Ergonomics of the thermal environment — Assessment of the influence of the thermal environment using subjective judgement scales
11. ISO 11399, Ergonomics of the thermal environment — Principles and application of relevant International Standards
12. ISO 12894, Ergonomics of the thermal environment — Medical supervision of individuals exposed to extreme hot or cold environments
13. ISO/TS 13732-2, Ergonomics of the thermal environment — Methods for the assessment of human responses to contact with surfaces — Part 2: Human contact with surfaces at moderate temperature
14. ISO/TS 14505-1, Ergonomics of the thermal environment — Evaluation of thermal environments in vehicles — Part 1: Principles and methods for assessment of thermal stress

(*Continued*)

TABLE 23.2
(Continued)

15. ISO 14505-2, Ergonomics of the thermal environment — Evaluation of thermal environments in vehicles — Part 2: Determination of equivalent temperature
16. ISO 14505-3, Ergonomics of the thermal environment — Evaluation of thermal environments in vehicles — Part 3: Evaluation of thermal comfort using human subjects
17. ISO/TS 14415, Ergonomics of the thermal environment — Application of International Standards to people with special requirements
18. ISO/TS 15666, Acoustics — Assessment of noise annoyance by means of social and socio-acoustic surveys
19. EN 15251, Indoor environmental input parameters for design and assessment of energy performance of buildings addressing indoor air quality, thermal environment, lighting and acoustics
20. Boyce, P. R. (2003), Human Factors in Lighting. 2nd Edition, Taylor and Francis. ISBN 0-7484-0950-5
21. Griffin, M.J., (1990) Handbook of human vibration. Academic Press. ISBN: 0-12-303040-4
22. Parsons, K. (2003) Human Thermal Environments, 2nd Edition, Taylor and Francis. ISBN 0-415-23793-9
23. Sato, H., Morimoto, M. and Wada, M., Relationship between listening difficulty and acoustical objective measures in reverberant sound fields, Journal of the Acoustical Society of America 123(4), 2087-2093 (2008)
24. Wilson, J. and Corlett, N. (2005) Evaluation of Human Work. 3rd Edition, Taylor and Francis, ISBN 0-415-26757-9

TABLE 23.3
Definitions of environmental ergonomics methods from ISO 28802 (2012)

An adaptive opportunity is the opportunity for a person to alter the environment to which he or she is exposed by behavioural (move away, adjust posture, adjust clothing, etc.) or other means (e.g. open window, close door, adjust environmental controls).

A behavioural method is a method that quantifies or represents human behaviour in response to an environment.

An objective method is a method that quantifies the physical, physiological or psychological condition of a person by the use of instrumentation or measures of output such as performance measures.

A subjective method is a method that quantifies the responses of people to an environment using subjective scales.

DEFINITIONS

The standard provides the following definitions (Table 23.3).

Guidance for the design of an environmental survey includes the aim, measurement of the physical environment and subjective responses, where, what and when to measure, who to include in the survey and adaptive opportunities. Chapter 24 provides a case study involving the assessment of a total environment in an office using an environmental ergonomics survey.

FURTHER READING

Parsons, K. C., 2015, The environmental ergonomics survey, In Wilson, J. R. and Sharples, S., *Evaluation of human work* (4th Ed), CRC Press, Taylor & Francis Group, Boca Ratan, New York, Oxford, ISBN 978-1-4665-5961-5, Chapter 23, pp 641–654.

STANDARDS

ISO 28002, 2012, *Ergonomics of the physical environment -The assessment of environments by means of an environmental survey involving physical measurements of the environment and subjective responses of people*. ISO, Geneva.

24 Case study
Environmental Ergonomics assessment of a high-quality 'gold standard' office

THE ENQUIRY

Joshua had been given a specific task by the chief executive of the company, to ensure that the newly formed 'high-flyer' office in pleasant surroundings in the headquarters building on the west coast was second to none in terms of environmental quality.

Daisylu studios was a state-of-the-art visual production and software company that had (secretly) moved into producing new movies from (often modified) clips and scenes of existing movies. They used artificial intelligence (AI) and identified additional scenes that glued the movies together. A demonstrator movie had been produced (highly confidential) that was deemed 'fantastic' and at relatively low cost, so the company was convinced that this was the future and that a completely acted out new movie was a thing of the past.

Their first task was to produce a series of 'new' movies derived from a well-known spy thriller collection. The use of a movie series was particularly appropriate as the same main character and actors provided movie scenes that could be adapted for the new movies.

An organisational structure based upon skills required (software, movie production, legal and accounting services and management) was constructed and no expense was spared to attract the best people in the world. As well as salary, resources and flattery, people were attracted by others in the team, the location and the working environment. Script writers from around the world were given themes, but they were not informed about how their script would be produced into a movie. Databases of movie scenes with details of actors, locations, dialogue and more were linked to artificial intelligence software that could match scripts to collections of scenes to provide the starting point for the new movie.

Joshua was responsible for maintaining satisfaction among the workers to facilitate production and had an extensive budget and support team including skilled computer technicians. The office was designed to provide a productive and futuristic atmosphere in an attempt to provide inspiration as well as to promote a feeling of being valued and encouraged. The layout was traditional as Joshua wanted people to realise that they were there to work and not play.

On researching the problem Joshua decided that an environmental ergonomics post-occupancy survey would re-assure the office workers that they were being valued and that he would be able to confirm that the office environment met the highest

 DOI: 10.1201/9781003401964-32

environmental standards for health, comfort and performance and more. To complement the ethos of the company, he identified what appeared to be a 'high-flyer' company, with personnel of international reputation for excellence, called 'Double E' and contacted Wally, its managing director, as follows.

Dear Professor,

I am writing to enquire whether you would be able to conduct an environmental ergonomics assessment of what we hope is, and aim for it to be, an office of the highest quality that we have constructed to facilitate production by selected workers on a new initiative for our company.

The workers have been working in the office for three months and with only a few 'teething' problems, we think that it is as we hoped.

I would like a full assessment of the total office environment and a report that hopefully will confirm that we have created a successful and productive environment, that I can send to my senior management. Any recommendations of course would be most gratefully accepted and acted upon.

Yours sincerely

Joshua

Manager of special projects, Daisylu studios

DOUBLE E

Wally was professor of environmental ergonomics at the prestigious Minerva University, known for its outreach and enterprise activities. He had set up the company with his four successful PhD students (Drs, Andrew; Miranda; Lulu: and Peter) and it had been operating for one year. The company was set up under a 'start up' scheme, an independent company of limited liability but with university support. The university provided administrative, legal, contractual and insurance support. When acting as consultants, each person had their own personal liability insurance through their ergonomics professional body.

Andrew had completed an undergraduate degree in civil engineering and a Master of Science (MSc.) in Ergonomics at Minerva University where he completed a six-month project into noise annoyance in buildings. He went on to successfully complete a PhD into human response to whole-body vibration discomfort under the supervision of Wally.

Miranda had completed an undergraduate degree in optics and a Master of Science (MSc.) in Ergonomics at Minerva University where she completed a six-month project into Heat stress and protective clothing. She went on to successfully complete a PhD into Lighting in lecture rooms under the supervision of Wally.

Lulu had completed an undergraduate degree in human biology and a Master of Science (MSc) by research in Ergonomics at Minerva University where she completed a twelve-month project into physiological measurement of responses to environmental stress. She went on to successfully complete a PhD into thermal comfort in buildings under the supervision of Wally.

Peter had completed an undergraduate degree in human psychology and a Master of Science (MSc.) in Ergonomics at Minerva University where he completed a six-month project into subjective assessment of air quality. He went on to successfully complete a PhD into Post-occupancy questionnaires for offices under the supervision of Wally.

All members of the team had successfully completed theoretical and practical modules (classes) into environmental ergonomics as part of their MSc. Wally was module leader and experts in particular environmental components made contributions, with practical classes in the environmental ergonomics laboratories as well as conducting 'field' work in groups.

THE CONTRACT

On receipt of the letter, Wally discussed the project with Joshua over the telephone. He indicated that the work could be undertaken but was Joshua sure that he wanted to go ahead. It seemed possible that after an extensive (and expensive) environmental ergonomics assessment, it could be concluded that the environment was of high quality and that no further work was required. Maybe a simple assessment could be made to confirm that there were no major problems.

Joshua confirmed that there were no complaints about the office environment and in any case he had a rapid response policy to any request from the workers. He required the full survey that included a comparison between the office environment and what would be internationally accepted as a gold standard office environment.

The team of architects, interior designers and building service engineers that created the office had been among the best available. As well as their extensive expertise, they had involved users in the design through focus groups and video conferencing. They had provided a full report that would be made available to Wally.

Joshua wanted an independent post-occupancy evaluation, to maintain the 'valued – no expense spared' ethos among the workers as well as for himself and as evidence he can provide to senior managers that he is achieving his objectives.

An important requirement was that the integrity of the work of the office was maintained. Joshua, the workers in the office and all of his team had signed non-disclosure agreements and in any case information was on a 'need to know basis'. Wally, his team and anyone else involved through 'Double E' would be required to sign a non-disclosure agreement. They would not discuss the work with the workers or anyone else and concentrate on the environmental ergonomics assessment.

AGREED AIM AND OBJECTIVES

Wally agreed that they could go ahead and would provide a contract, timing and costs. The following aim and objectives were agreed in writing by email.

Aim: To carry out a post-occupancy environmental ergonomics evaluation of an intended high quality office environment (title and location to be provided).

Objective 1: To make measurements of the physical environment in the office
Objective 2: To determine the worker response to the environment in the office
Objective 3: To compare measurements of the physical environment in the office with accepted 'gold-standard' conditions
Objective 4: To suggest any improvements if required

Objective 5: To provide five copies of a 'commercial in confidence' report exclusively to Joshua within one month of the completion of the survey. No other copies to be kept and certainly no publications relating to the work, in the academic literature, the press or elsewhere.

After the telephone conversation, Wally met with all of the Double E team to explain that this was an opportunity for experience and resources. Some discussion was had over the non-disclosure agreements, mainly because as academics they were not comfortable with 'secrecy'. Some had had experience with the official secrets act where signature means a life-time agreement for the activity under consideration but also for any other information. This was an unacceptable restriction of academic freedom. Wally reminded them that this was not an academic exercise but that he will ensure that the non-disclosure agreements were minimal and directly relevant to the project.

After some negotiation all administration was completed and the project went ahead. During the negotiations, it was agreed that the scope of the survey would include workstation environments but not workstation assessment. It was also decided that the 'Double E' team would not meet the office workers and that all measurements would be taken at a distance.

ACHIEVING OBJECTIVES

The Environmental Ergonomics Survey

The 'Double E' team agreed to work in pairs to cover the environmental components: thermal (Lulu and Miranda); acoustic (Andrew and Wally); vibration (Wally and Andrew); light (Miranda and Peter); and air quality (Peter and Lulu); with first named as the lead. A full team meeting would be held when all results were obtained, and conclusions made, to consider the integrated environment as a whole. Interim reports for each environmental component would be presented to all in good time before the meeting. Wally would closely manage all activities and would liaise with Joshua.

The Office

A plan of the office with full building services was provided to the 'team'. The office was rectangular and had dimensions 30m × 10m with a high ceiling of 4m and desks placed well apart in a horseshoe configuration facing the windows and with a replica Henry Moore abstract Statue at the focal point, on a plinth and about 2m tall.

Each consisted of computers and large screens, as well as high-end chairs. The thermal environment was controlled by a displacement ventilation system using seven cool air 'bins', under the plinth and evenly spaced at the bottom of the walls. The windows were sealed, double glazed and along one north facing wall. Lighting was mainly daylight during the day, supplemented to full artificial light when dark outside. The building was in a quiet area with minimum external noise and no apparent source of vibration.

There was a kitchen and a separate labelled conference/screening room that were not part of the survey. Building services were over-specified for the room allowing for a wide range of environments and excellent control. Ventilation systems allowed for well over the required number of air changes per hour. Automatic blinds that could also be manually adjusted, were provided on the windows. The room was mainly of pastel grey colour with posters of movies on the walls providing both inspiration and purpose.

The 24 workers (14 male, 10 female, age range 25 to 60 yrs) worked in the office from 9.0 am to often late in the evening, from Monday to Friday. Each had their own workstations at fixed locations. They were not allowed access at weekends when extensive cleaning took place on Saturdays and full technical maintenance on Sundays.

Workstations did not have 'public' internet access but each worker had a separate computer workstation on a long desk at the back of the office. This was used for internet access and communications.

The Meeting

The whole team flew to the west coast to meet with Joshua and relevant support workers in the headquarters building one Saturday. Wally explained that however 'high tech' the measurement, transmission and recording system was, the quality of the data, that represented the physical environment, will initially depend upon the location and transducer system and the laws of physics. Transmitting data from a light sensor placed behind a screen will not be a valid representation of the visual environment experienced by workers. Temperature sensors on the wall do not represent worker experience etc.

It was agreed that a trolley system would be used and that measurement sites should be between workstations for the thermal, acoustic and lighting environments. That is 15 trolley systems in all. Accelerometers would be fixed to the floor at six sites evenly spaced between workstations to record vibration (see Figure 24.1. Two trolleys were evenly space across the rear long desk containing the 24 communication computers.

Measurements were transmitted to 'Double E' and were continuously monitored and recorded as absolute values with time. Further analysis produced Leq,dB(A) values over 8 hours; Predicted Mean Vote and Predicted Percentage Dissatisfied values (PMV/PPD); illuminance levels in lux (lx); and vibration weighted rms acceleration levels and dose values.

Measurement of air quality was mainly performed using subjective judgements of the workers. However, it was agreed that 12 female office workers from another part of the building would wait outside in the fresh air, each day just before the closure of work. They would quickly enter the room, take a position next to a designated trolley, and smell the air. They would make an instant judgement of whether the air quality would be acceptable or not if they had to work at that workstation.

They would report back to a worker in the outside room, not receive knowledge of previous ratings and not discuss the office with their colleagues and not communicate with the workers or anyone else while in the office. They would be paid for their

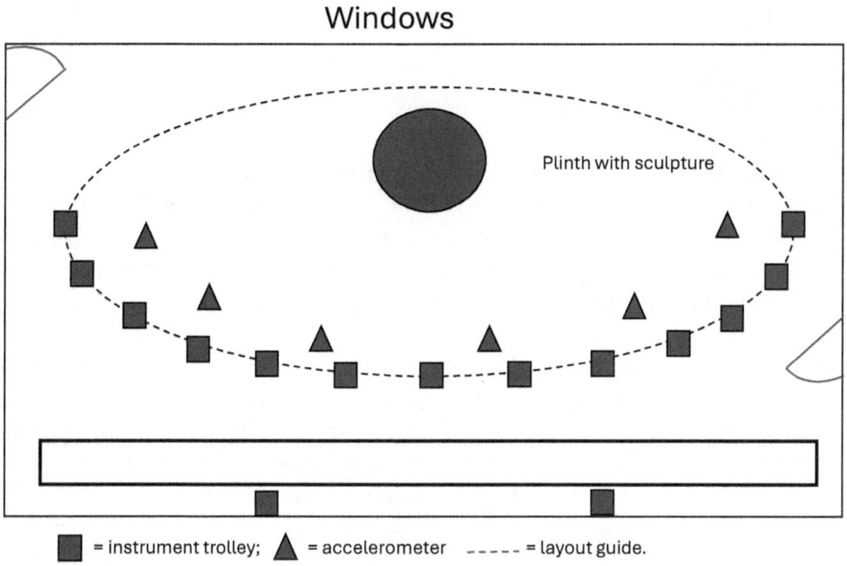

FIGURE 24.1 Measurement sites to determine if the office can be considered as 'gold standard'.

cooperation. 'Not acceptable' responses were assumed to indicate dissatisfaction and were used to estimate decipol values (level of pollution).

The equipment was set up on Sunday with the assistance of the excellent computer support team. It was agreed that Wally would remain next to the office for the full week of the survey, with instruments operating between 8.0 am and 6.0pm Monday to Friday. He would check instruments, monitor data and take independent recordings as a backup. The rest of the team flew back to 'Double E' headquarters and tested the signals from the office as well as receiving a completed test questionnaire.

The Pilot Trial

Before measurements were taken in the office, a pilot trial had been conducted to ensure that anticipated measurements and procedures would deliver the desired outcome. A room was hired from the university and was 'converted' into a functional simulation of the office environment to be investigated, including computer workstations.

Office workers from the university were used as participants in the trial and instruments were set up to measure the environment. Complete data were collected, analysed and discussed by the team until it was concluded that all objectives could be achieved. The data collection periods were over two evening sessions.

As well as being present for the measurements of the environment in the simulated office' the participants completed the proposed subjective questionnaire as it would be presented to the workers. This was to ensure that it was understandable and that it would provide the required information. Revisions were made until it met the requirements.

Objective 1: To Make Measurements of the Physical Environment in the Office

The instruments used for measuring the thermal environment were according to the specification given in ISO 7726 (1998) for air temperature; radiant temperature; air velocity and humidity (B&K climate analyser type 1213 (Olesen, 1985), modified to give direct readings to the internet) as well as a black globe thermometer 150mm diameter, with a thermistor at its centre to give globe temperature and derive mean radiant temperature when corrected for air temperature and air velocity.

All equipment was set up on the trolleys at 1m height and was unrestricted. Air temperature and air velocity were also measures at a height of 1.5m and 0.2m. Clothing worn and activity level were observed via the overhead camera and values estimated using ISO 9920 (2007), and ISO 8996 (2004), respectively. Noise levels were measured using B&K sound level meter type 732A, that gave dB(A) and dB(lin) levels continuously throughout the day (IEC 61672-1, 2013). Lighting levels were measured using colour corrected light meter type Minolta T10A, to provide horizontal illuminance (lux).

The measurement systems were positioned before workers entered the office on Monday. They did not interfere with work and were correctly explained as systems that were to ensure that the best possible office environment was provided and that they would be there for one week.

All instruments were calibrated and certified by national physics laboratories. They were also tested and compared in the Minerva university environmental ergonomics laboratory environmental chambers, where the trolleys were set up and measurements recorded. So absolute values as well as relative values across transducers and measurement system were taken, and replaced where necessary until validity and consistency were achieved.

Accelerometers were tested in the laboratory, but also turned upside down just before installation, to determine a voltage equivalent to the acceleration due to gravity (9.81 ms^{-2}). Sound level meters were calibrated using pistonphones placed over microphones giving a standard pure tone of 94 dB (A and lin) at 1000 Hz. Independent measures were used such as the video recording of bubbles to compare with the anemometer readings and the use of a whirling hygrometer to give an individual comparison for air temperature and humidity.

Objective 2: To Determine the Worker Response to the Environment in the Office

A wide-angle lens camera was already in place at the centre of the ceiling such that the whole office could be viewed and activity recorded. This live feed could be sent to ‘Double E’ via the internet for live observation and recording. It would be used for monitoring any problems and interference with equipment as well as for any behavioural and adaptive responses of workers. It was also used to estimate metabolic heat production due to activity and to estimate clothing insulation values.

Although usually taken in the environment under assessment, the video recording was used to complete an observation checklist by Lulu and Miranda. The question of whether they would like to work in that office was a clear yes. The general impression was of a stimulating environment with little distraction from work and that adaptive opportunities allowed adjustment of clothing. The requirement of 'smart casual' set the tone and was adhered to and loosely monitored. This was as well as ability to move around, and of control of blinds, to open and close windows on request to ensure that it complemented other systems.

Subjective measures were taken from individual workers who were identified by their workstation. Workers were informed that this was not a test, but that honest personal opinions were required. At exactly 11.0 am and 3.0 pm a questionnaire would appear on screens, for immediate completion. The questionnaire covered questions about the whole environment and took about 10 to 15 minutes to complete. It was an online questionnaire and the responses were immediately sent to 'Double E' for recording and analysis (see Appendix 12).

Objective 3: To Compare Measurements of the Physical Environment in the Office with Accepted 'Gold-standard' Conditions

Standards and guidelines are usually created to preserve health and functionality and optimum environments to provide comfort and satisfaction. Environments that provide inspiration, delight, pleasure and more merge with organisational factors and relate to the dispositions of the occupants of the space and of the 'total system'.

This environmental ergonomics survey has as its scope to compare the physical environment with what could be regarded as the best possible. In the first instance this could be to provide comfort and satisfaction (avoid discomfort and dissatisfaction). It can be assumed that comfort conditions also provide optimum environments for performance as the RDC method (Parsons, 2018; ISO TR 23454, 2019) would provide time available after limits have been taken off due to regulations = 100% (1) × time on task after time off task due to distraction had been taken off = 100% (1) × capacity at conducting tasks after reduction due to discomfort and stress have been taken off = 100%; so expected performance = 1 × 1 × 1 = 1 = 100% and no loss due to the physical environment and a basis for providing optimum production.

The thermal comfort standards ISO 7730 (2024) and ANSI/ASHRAE 55 (2023) as well as the energy standard EN 15251 (2007) provide gold, silver, and bronze conditions for thermal comfort, These are based upon the range of PMV/PPD (Predicted Mean Vote/Predicted Percentage Dissatisfied) values and for a gold standard this is PMV±0.5, so PPD<6%. For the office environment under consideration this was modified to achieving PMV=0.0 with no dissatisfaction and adaptive opportunity available for each occupant to achieve the gold standard. A similar approach was taken for other environmental components and their integration.

RESULTS OF THE SURVEY

Physical Measurements

Objective 1: To make measurements of the physical environment in the office

Data recording began at 8.0 am on Monday morning and ended at 6.0 pm on Friday. The period between 8.0 am and when workers entered the office at 9.0 am and 5.0 pm, when workers nominally left the office, and 6.0 pm was observed and used to provide background information. It was not used in the main analysis.

In analysis, measurements were integrated to representative values over five-minute periods throughout the day. In addition to the traditional environmental measures, the maximum and minimum values for each environmental component were made as well as a representative daily average value. The values at 11.0 am and 3.0 pm were analysed in detail as these were the times when the subjective assessment was made.

The thermal environment was quantified by integrating values of air temperature, mean radiant temperature, air velocity and humidity along with estimates of clothing insulation and metabolic rate to provide PMV values on the scale: +3,hot;+2,warm;+1 slightly warm;0,neutral;-1,slightly cool; -2,cool;-3,cold. Values obtained ranged from -0.5 to + 0.8 across the workstations and throughout the day with an average of 0.3. Over all workstations, at 11.am the PMV was between -0.4 and -0.1, and at 3.0 pm it was between 0.5 and 0.7. Consideration of local thermal discomfort suggested possible dissatisfaction due to draughts in the morning, when air velocities were recorded at a maximum of 0.25 ms^{-1} at the feet at two of the workstations near ventilation cool air bins. Values of other local thermal discomfort indicators such as asymmetric thermal radiation, vertical air temperature gradients etc, did not indicate that they would cause discomfort (see ISO 7730, 2024). Table 24.1 provides a summary of results for each of the five days.

Noise is unwanted sound so we must judge if the sound of computers in the office is unwanted. Without the computers switched on and in the morning before the start of work, the average sound level was around 45 dB(A) and 47 dB(lin). During work

TABLE 24.1
Measurements of the thermal environment

	Monday	Tuesday	Wednesday	Thursday	Friday	Mean
Predicted Mean vote (PMV)						
+3,hot; +2,warm; +1,slightly warm; 0, neutral; -1,slightly cool; -2,cool; -3,cold						
11.0am	-0.3	-0.4	0.0	-0.1	-0.4	-0.24
3.0pm	0.6	0.6	0.5	0.7	0.8	0.64
Daily average	0.2	0.3	0.0	0.5	0.4	0.28
Predicted percentage of dissatisfied people (PPD)						
PMV = 0, PPD = 5%; ±0.5 = 10%; ±1 = 25%; ±2 = 75%; ±3 = 100%						
11.0am	7%	8%	5%	5%	8%	6%
3.0pm	13%	13%	10%	15%	18%	14%
Daily average	6%	7%	5%	10%	8%	7%

with computers switched on, the maximum value was 55 dB(A), and minimum value 50 dB(A) with an equivalent energy value of 52 dB(A) Leq over all 5 days. At 11.0 am the average sound level was 53 dB(A) and 58 dB(lin) over the five days and at 3.0 pm it was 55 dB(A) and 60 dB(lin). dB(lin) values were greater due to low frequency fan noise of the computers.

Vibration levels were surprisingly high and predicted to be perceptible mainly caused by fans from computers that were placed on the floor next to each workstation. Between 9.0 am and 5.0 pm the weighted vibration level (ISO 2631-2, 1989) was 0.02 ms^{-2} rms; with a maximum value of 0.04 ms^{-2} wrms and a minimum value of 0.005 ms^{-2} wrms.

Although facing North (Northern hemisphere), the weather outside was consistent sunshine and most of the room was lit by daylight. This dominated the visual environment as workstations were orientated towards the windows with the plinth at the focal point of an approximate semi-circle of workstations. The computer screens had their own backlighting and supplementary lighting directly above each workstation, apparently providing no reflections, glare or flicker, when daylight reduced, was provided for evening work. For working without daylight, automatic blinds closed and fluorescent lighting was available that provided an even distribution of light across the room. The horizontal illuminance levels on each trolley, paced adjacent to workstations, was 1,243 lux on average over all measurement sites and across the whole week (daytime), with a minimum value of 800 lux and a maximum value of 2000 lux.

Air quality was judged by the 12 female judges as acceptable or not acceptable at the location of each measurement trolley at the end of each day with workers present. Out of the 5 × 12 = 60 judgements, 2 of the judgements rated not acceptable, interpreted as 3% dissatisfied and 0.2 decipol. Comments were made that there was an 'electrical' smell from computers that most said that was acceptable but the two 'not acceptable' ratings related to that.

WORKER RESPONSES

Objective 2: To determine the worker response to the environment in the office

Although 24 workers × 2 sessions (11.0 am and 3.0 pm) × 5 days = 240 sets of ratings and comments were taken, to avoid repetition, general findings with specific points are provided below. In summary all workers were very satisfied with their working environment both 'NOW' and 'GENERALLY AT WORK'.

For the thermal environment, on average and over all measurements, workers said that they felt, neutral, not uncomfortable, not dry, not draughty, found the thermal conditions acceptable and pleasant and that they were satisfied with the thermal conditions and wanted no change. They all felt that the environment encouraged productivity and did not interfere with their ability to perform their tasks. Two of the female workers suggested that they felt a cool draught in some areas and along with two other female workers indicated that they would like to be slightly warmer. 220 of the 240 ratings, indicated that they would like 'no change'.

In general workers found the acoustic environment not noisy, not annoying, acceptable and that they were not dissatisfied. It was not regarded as pleasant but comments indicated that it was the nature of the job and that maybe the buzz of computers suggested a productive environment. The absence of computer sounds might be worse!

All workers indicated that the vibration in the room was acceptable. In general it was perceptible, not annoying, not distracting, not pleasant but it required no change. No additional comments were made.

In general the visual environment was found acceptable, bright, pleasant to work in and suited to the task and ethos of the office with acceptable colour scheme. However it was the area where most comments were made, not of concern but with opinions and suggestions. Workers found the lighting levels not uncomfortable, acceptable, not dim and pleasant with no glare, flicker or interference with work, particularly on screens. 12 of the 24 workers suggested that the plinth and sculpture were cast in shadow with bright light behind and it blocked the view. Framed posters on the walls were generally appreciated although six of the workers commented on the reflection of windows in the glass covering the posters.

All workers indicated that the environment was not smelly, that ventilation appeared adequate, did not think that the environment gave them headaches and thought that the air quality was acceptable.

Is the office 'gold-standard'?

Objective 3: To compare measures of the physical environment in the office with accepted 'gold-standard' conditions.

Gold standard conditions for thermal comfort require a PMV value within ±0.5 with no local thermal discomfort (ISO 7730, 2024). That is between neutral and 'halfway' to slightly cool or slightly warm as a predicted average.

Observations of the behavioural responses of workers, including moving around and adjustments to clothing strongly suggest that this is achieved if adaptive opportunities are considered. In fact, each person will be able to achieve PMV = O, neutral (comfortable), and no dissatisfaction. For local discomfort, Predicted dissatisfied due to draught will be 40 % for 23 °C air temperature and air velocity 0.25 ms^{-1} (maximum measured). This may require some attention for a gold standard but will also be ameliorated by adaptive opportunity.

For the acoustic environment, levels of conversation in a room can be up to 65 dB(A) and annoyance is often related to speech interference. Levels of a countryside at night (20 dB(A)); quiet office (40 dB(A)); busy computer office (55 dB(A)); and city traffic (80 dB(A)) are examples of sounds that may cause annoyance and loss in performance but will not cause harm. Values less than 40 dB(A) could be taken as gold-standard guidelines for offices. The noise levels in the office could not then be considered as gold-standard in terms of guidelines.

Absolute vibration levels are around 0.01 ms^{-2} rms and limits for an office during the day for continuous vibration will be around 0.04 ms^{-2} rms (ISO 2631-2, 1989). It is reasonable to assume that a gold-standard would be at levels below the absolute perception level. Maximum levels measured in the office were 0.04 ms^{-2} rms so are above what could be considered a gold standard.

It is universally found that people prefer daylight to artificial light and as such daylight could be regarded as a starting position for a gold-standard visual environment.

It provides both brightness and interest, with variation internally and also a view (see Boyce, 2014).

Direct sunlight can provide glare but will not be experienced in north facing windows (northern hemisphere). Daylight is then derived from the sky and is termed skylight. To be entirely satisfactory a window area should be greater than 30 % of the total wall area, with an inspiring unrestricted view. Recommendation for horizontal illuminance in offices vary internationally but 500 lux would generally be regarded as good lighting and above 1000 lux would be regarded as very good lighting . Levels in the office therefore would be regarded as towards a gold-standard.

The equivalent for air quality, to lux for light and dB(A) levels for sound, is the decipol. This is a measure of the pollution level in the office and depends upon the strength of pollution sources (bioeffluents for 24 workers, chemicals from office materials etc) and the ventilation rate. For a percentage dissatisfied of 3% the perceived pollution rate is around 0.2 decipols. CEN CR 1752 (1998); gives 3 categories of air quality. Category A,B and C corresponding to 15% ; 20% and 30% dissatisfied and high, medium and moderate level of expectation respectively. ANSI/ASHRAE 62.2, 2022, suggests an acceptable air quality provides 20% dissatisfied or less.

Decipol level (Ci) is given by the equation $Ci = 112(\ln(PD) - 5.98)^{-4}$. For category A, the highest standard a decipol level would be based upon PD = 15%. So for PD=15% gives 1.0 decipol. PD=20% gives 1.4 decipol; PD=30% gives 2.5 decipol. An accepted limit is often taken as 1.5 decipols.

For the office environment, 3% dissatisfied corresponds to 0.2 decipol and well within limits. It would be reasonable to conclude a gold-standard had been reached for air quality. However, for improved air quality lower air temperatures may provide fresh air but maybe at the expense of thermal comfort.

THE INTEGRATED ENVIRONMENT

Team Meeting to Integrate Results

After all analysis was complete, a summary report with conclusions for each environmental component was prepared by each of the teams and presented to a meeting of the whole team, by the team lead. After some discussion the following was agreed.

1. The thermal environment provided gold-standard conditions and when adaptive opportunities were included, the environment can be considered to be of the highest standard.

   No further action proposed in general. Although further consideration given to comments by four female workers.
2. Noise levels in the office were acceptable but may not be of gold-standard. Worker responses did not consider noise to be a problem and some indicated that the sound of computers is an indication of activity. Despite the lack of concern among workers it should be remembered that noise reduction usually improves an environment, and it is unlikely that workers will complain if computer noise were reduced or eliminated.

   Action to reduce computer noise.

3. Vibration was at perceptible levels but was not considered a problem by workers. If it were reduced by removing computers it is not expected to cause a problem and should reduce vibration to imperceptible levels.
   Action to reduce computer vibration.
4. Illuminance levels were commensurate with a high-quality office environment. However, the visual environment was clearly of importance to the workers and provides an opportunity for improvement. The plinth and statue seem to be appreciated but also spoil the view for some workers to the outside through the windows. It lacks definition due to disability glare when viewed from some workstations. The glass within the frames of the movie posters on the wall should be removed or replaced to eliminate reflections.
   Action to further consider the visual environment, particularly frames, posters and statue.
5. Air quality is of a high standard. Improvement could probably be made by reducing air temperature but that will reduce thermal comfort or require unacceptable levels of clothing insulation.
   No action required.

NOTE ON TRANSIENT EFFECTS AND CONTROL

When considering environments that are welcoming, provide pleasure and project 'high-quality', transient effects should be considered. Moving from an undesirable environment to a desirable environment can elicit pleasure.

For example, it is a well known tactic (as well as good design) to provide air conditioned cool air when entering shops or hotel entrances from hot climates, or warm conditions when moving from cold outside conditions. Similar responses can be obtained by moving from noisy to quiet (or quiet to loud music venues to create excitement), and no doubt smelly to fresh air, vibrating vehicles to solid land and boring to interesting or bright glaring to more even lighting views can all promote pleasure.

On occasions the initial 'new' climate will not be appropriate for long term comfort. For engineering reasons as well as initial responses buildings are often overheated. A common psychological phenomenon is that people like to feel that they 'master' the environment. If it is hot outside they demonstrate that they can counter that with cold conditions. Not to be recommended for comfort or energy efficiency, but a real phenomenon.

For the present 'gold-standard' office it is recommended that, unless it becomes an issue, no action on transient effects is taken.

SUGGESTED IMPROVEMENTS

Objective 4: To suggest any improvements if required.

The results of the environmental ergonomics survey are that the office is of 'gold-standard' as required and it would be valid to report that result to workers and senior management.

There is an argument for leaving well alone, especially if it was felt that optimum performance was being achieved.

Considerations for improvement are

1. To modify computers to reduce noise and vibration levels.
2. To remove computers from the office so that only screens are present in the workstations.
3. To maintain levels of lighting and the visual environment specification.

   However to also consider how the ambiance and feel for the office can be maintained to complement its status as a futuristic, high-flyer, productive environment where the workers are valued.

   In environmental ergonomics this would imply involving users in design, with particular emphasis on dynamic change to avoid environmental boredom.

THE REPORT

Objective 5: To provide five copies of a 'commercial in confidence' report exclusively to Joshua within one month of the completion of the survey. No other copies to be kept and certainly no publications relating to the work, in the academic literature or elsewhere.

Five copies of the 'commercial in confidence' report were sent to Joshua as requested. The report contained full detail including raw data and the following executive summary.

EXECUTIVE SUMMARY

An environmental ergonomics survey was carried out on the total environment of office (location included). The objective was to evaluate the existing working environment using physical measures of the environment and to compare them with values that would be considered to be gold-standard' and an office environment of the highest quality. It was also to measure the workers experiences of the environment and how they felt about the environment when at work and to make recommendations for improvement if necessary.

All measurements were made from Monday to Friday, mostly by electronic transfer without the investigators coming into contact or interfering with the work of the 24 workers in the office.

Measurements were made of air temperature, radiant temperature, air velocity, humidity, dB(A) levels, illuminance levels (Lux), vertical vibration on the floor (weighted rms acceleration) and decipol levels to assess air quality. Comparison with what would be values for a high quality office environment showed that all measures could be regarded as meeting that requirement. Importantly there were few or no levels of dissatisfaction with the environment expressed by workers.

It is recommended that consideration should be given to leave well alone as a high quality environment had been achieved with which workers were satisfied. This is the conclusion of the environmental ergonomics survey. Of the issues raised, the interaction between workstations and cool air bins could be reviewed; some advantage could be gained by removing computers from the office leaving workstations with quieter conditions and no source of vibration; and although the visual environment was of high quality and acceptable, its importance was apparent and opportunities for enhancing the ethos of the office using the visual environment, should be explored.

OUTCOMES

Joshua was pleased to receive confirmation that the office was of high quality and that workers were satisfied with the environment. As far as the Double E team were aware the report was accepted and it is possible that recommendations were pursued. Actually the computers were removed to outside of the office and noise and vibration was reduced. Workers recognised the improvement.

The working environment was not the main concern of the workers who were highly motivated to succeed. A series on new movies was developed and produced. However, although received with acclaim, by audiences of the public used to gauge reaction, strikes among the acting and movie producing fraternity and legal constraints stopped the project. The office and some of the workers were moved to a country with a massive population and where copyright laws were essentially ignored.

The workers at Double E had conducted an environmental ergonomics assessment of an integrated environment and were satisfied that a good job, well done, had been achieved. Consideration of what could be achieved in an environment of the highest standard, was an opportunity to assess environmental ergonomics philosophy and methods and to consider limits. It was clear that health, comfort and performance were the bread and butter of environmental ergonomics and provide basic criteria for assessing environments.

Human characteristics however consist of more than functionality and the design of environments that provide stimulation, inspiration, pleasure, delight and more are a challenge for the future that will require art, science, inspiration and political, social and psychological awareness as well as physical measure and specifications.

FURTHER READING

Parsons, K. C., 2015, The environmental ergonomics survey, In Wilson, J. R. and Sharples, S., *Evaluation of human work* (4th Ed), CRC Press, Taylor & Francis Group, Boca Ratan, New York, Oxford, ISBN 978-1-4665-5961-5, Chapter 23, pp 641–654.

STANDARDS

ISO 28002, 2012, *Ergonomics of the physical environment -The assessment of environments by means of an environmental survey involving physical measurements of the environment and subjective responses of people*. ISO, Geneva.

Appendix 1

GENERIC INFORMATION REGARDING ENVIRONMENTAL ERGONOMICS SURVEYS AND CASE STUDIES

For each of the environmental components (heat, cold and thermal comfort; noise; whole-body and hand-transmitted vibration; light and air quality; as well as integrated total environments involving all components), a case study is provided. They are intended to demonstrate the practical application of knowledge of human response to the environment (environmental ergonomics). The case studies are presented in narrative form with technical content, using hypothetical first names (not representing real persons, etc.) to give a flavour for the nature and context of the task.

Implicit in all case studies are points that are relevant but not always recognised. The notes below provide general background information that will be useful to the environmental ergonomist embarking on a survey.

1. The client contacts an organisation (e.g. university research laboratory) and environmental ergonomist of respected and acknowledged reputation not only for the best available advice but also to establish credibility with peers as well as in any enquiry and litigation that may become involved.
2. The political context of any request can be important. The person making the request for advice may have been instructed to do so by a senior manager and hence may not have ownership of the idea or full commitment to the project. An outside expert may be consulted as an agreed neutral in an area of conflict within an organisation and so on.
3. A proposal with stated aims and objectives, outputs (deliverables) and costs would normally be sent to the client for approval. Security arrangements would be agreed, usually at the level of 'commercial in confidence'. Liability insurance for the consultant is usually required and is often provided through a limited company (e.g. Oxbridge University Consultants Ltd.) who will have standard legal contracts.
4. In environmental ergonomics, the term *stress* is used to describe the environment occupied by a person and *strain* to describe the human response to the environment.
5. It is fundamental to all surveys involving people that a person may withdraw from an environment or investigation without explanation and in a social environment that provides no pressure to continue. A request for a person to continue in an environment, when asked to withdraw, shall be ignored (irrespective of their position or rank) and they should be withdrawn to a safe environment if previously agreed (objective) criteria are met.

6. Participants include all involved in the project who are exposed to the environment, including 'subjects' as well as those involved in monitoring, carrying out and supervising the project.
7. It is essential that the aims and objectives of the project are unanimously and unambiguously agreed in writing and that any deliverables are clearly stated in writing with a clear understanding and in detail. It is also essential that dates and times are agreed for access to resources and for the report and other deliverables, as well as an address and person to whom they will be delivered and in what form.
8. Always do a good job, don't take on impossible tasks and refer to others if it is outside of your expertise. Safety is of the highest priority. Avoid unacceptable risks.
9. Once the contract has been agreed then, irrespective of the costs involved, the best quality service should be provided. Reputation comes above all. Take time to ensure that the contract is clear and costed correctly, and don't take on a project that may end in a poor quality survey.
10. Take time to ensure that you and your team of investigators are fully acquainted with the terms and vocabulary of the people involved with the environment. This includes the names and geography of the environment to be assessed as well as any local and specialist terms used by the workers and management.
11. Screen the investigator who conducts the observation checklist to ensure that any characteristics (e.g.visual colour defects; hearing impairment etc) do not bias the results.
12. Be aware of the 'Hawthorne effect' where people try to please so are overly positive about the environment. Some occupants may use the survey as a displacement activity as they are more interested in the survey than their work.

Appendix 2

SCREENING OF PEOPLE BEFORE EXPOSURE TO PHYSICAL ENVIRONMENTS

Screening is a method for ensuring that people are not exposed to environments that will damage their health. It assesses the health of a person and compares their health status with contra-indicators that are known to cause vulnerability when exposed to the particular environment of interest. It is often a requirement for people to have a health check before they begin a new job.

It is important to recognise that screening is about the risk of exposing an individual to an environment. It is not about populations. For example, gender has been used as a contra-indication for preventing females to work in coalmines, firefighting or the army when some people will be entirely suited to that work. For people with disabilities, often excluded from environments, adaptive opportunities and environmental design may be able to avoid risk, and so on. That is not to say that unacceptable risk should be taken.

When I conducted research into heat stress in compressed air tunnels, as I was over 40 years of age, I was not allowed in the workplace. A screened younger researcher took my place. Although some people over the age of 40 years would not have an adverse reaction (and some younger would), it felt like a wise precaution and acknowledges that physiological systems generally deteriorate with age and that risk increases.

Screening can be part of risk assessment and design can reduce risk. It is also important to remember that it is an holistic assessment, supported by a checklist procedure, but with the final recommendation made by an identified expert and qualified healthcare professional.

There are general contra-indicators that make a person vulnerable to all stressors and those specific to particular environmental components. A selection of both is provided below.

GENERAL SCREENING

In some countries, a health service identification number will give qualified professionals access to medical records. In others, privacy laws prevent that. Where access is possible, details of the person to be screened can be obtained. In some countries, it is illegal for an employer to request medical records; however, they can use independent and confidential health services. It is often a statuary obligation to do so.

Identification details include: full name; health identification number; contact details for usual medical practitioner; gender; date of birth; and the individual's contact details. Medication should also be identified and checked against possible interactions with environmental exposure. General assessment may include: height;

weight; blood pressure; heart rate (and ECG); and fingertip oximeter readings. Urine and blood analysis will also reveal the general health condition.

Standard health check protocols are available as well as general health contra-indicators (for all environments). These include: a history of heart problems including atrial fibrillation; high blood pressure; obesity; diabetes; and more (see ISO 12894, 2001). Tests for musculoskeletal health, skin health, respiratory health, audiometry, visual performance and vibration perception are often recommendations. It is impossible to give a definitive list, which is why this must be an individual assessment and an holistic process.

Ethical issues are also of importance and a written statement of the environmental exposure with detail and a written and signed agreement by the person to be exposed, confirming that the exposure has been explained, is also required. No exposure shall take place if it is beyond limits for those fit for work. Actually, I would ask the question, do you think that you should be exposed to (the environment – (e.g. heat)? Even though there may be many reasons why the person would say yes and not be taken at face value. 'No' would be a clear contra-indication.

SCREENING FOR EXPOSURE TO HEAT AND COLD

For hot conditions (e.g. WBGT >25 °C), contra-indicators include diabetes myelitis (or kidney or bowel conditions); episodes of fainting; general imbalance and dizziness; disease of heart or blood vessels; high blood pressure; chest disease including asthma; mental illness including anxiety and depression; skin conditions; reduced ability to sweat (e.g. due to sympathectomy); medication that interacts with the effects of heat; and any reactions to heat and heat illnesses (provide details, especially of hospitalisation) (ISO 12894, 2001; Parsons, 2014).

For cold conditions (air temperature <0 °C), contra-indications include experience of fits, faints, collapse or loss of consciousness; thyroid disease; diabetes myelitis; general imbalance and dizziness; disease of heart or blood vessels; high blood pressure; Raynaud's phenomenon or other peripheral vascular disease; chest disease including asthma and chronic bronchitis; mental illness including anxiety and depression; skin conditions; rheumatism or disease of the joints, bone breakages, including metal pins; allergic reactions to cold; medication that interacts with the effects of cold; freezing or non-freezing cold injuries; and any reactions to cold and cold illnesses (provide details, especially of hospitalisation) (ISO 12894, 2001; Parsons, 2014).

SCREENING FOR EXPOSURE TO NOISE

The UK Control of Noise at Work Regulations, 2005 (HSE, 2005a), provides screening requirements in terms of employer and employees legal duties. A sample noise and health questionnaire includes medical details: Do you wear a hearing aid?; For both left and right ear, is hearing good/fair/poor?; Have you suffered injury/trauma to ears?; Have you suffered earache or ear discharge?; Is there ear disease of deafness in the family; Have (or do) you suffer from: head injury; ringing in the ears/head; dizziness/giddiness; exposure to ototoxic drugs or solvents; gunfire/blasts/explosions; and do you have any noisy hobbies (e.g. motor sports; ride a motorcycle; DIY; discos/loud music; shooting; other)?; Do you hear worse in noise?; and Have you had wax removed from your ears? Details are requested if answered YES. A final section is on previous noise exposure and whether hearing protection was worn as well as exposure time in years. Before starting work in the noisy environment, a baseline audiogram shall be taken.

Regulation 9 of the regulations concerns health surveillance and states that '... the employer shall ensure that ... employees are placed under suitable health surveillance, which shall include testing of their hearing'.

For auditory performance, screening may include measurement of performance on actual or simulated tasks. Audiograms will measure hearing capability and signal detection methods will provide useful information.

SCREENING FOR EXPOSURE TO VIBRATION

British Standard BS 7085 (1989) considers those unfit for experiments involving human response to whole-body vibration. These can be used for worker screening and include active disease of: the respiratory system; gastro-intestinal tract; genito-urinary system; cardiovascular system; musculo-skeletal system; and the nervous system, with detailed disorders given for each category. Pregnancy, mental health, recent surgical procedures and prosthesis shall also be included in screening. Griffin (1990) cites Seidel and Heide (1986) who suggest contra-indications as diseases of the vertebral column; scoliosis; significant fixed kyphosis, joint diseases; chronic diseases of the nervous system; psychotic disturbances; peptic and duodenal ulcers; chronic gastritis; diverticulosis; various veins; heamorrhoids; disorders of menstruation; prostatitis and people younger than 18 years of age.

SCREENING FOR EXPOSURE TO LIGHT AND TASKS REQUIRING VISUAL PERFORMANCE

Excessive or prolonged exposure to infra-red and ultra-violet light will cause tissue damage, and exposure should be avoided if possible and limited by exposure time if not. Interaction between a person and the lighting environment is dynamic and individual problems can often be overcome with design. Visual comfort and performance can often be assured using user trials so that screening may detect a potential problem that can be overcome. Boyce (2014) however describes 'special groups' for consideration and these can be considered as an integral part of the screening process.

The eyes of babies, and in particular premature babies, can be damaged by light and special consideration is urgently required. As the babies can't be screened, this must be a part of hospital practice. Other vulnerable groups include post-operative cataract patients and those with their lens removed. The effect of age will interfere with visual performance and is part of screening as is a history of eyestrain; migraine; autism; sensitivity to flicker and flashing lights, epilepsy, mental disorders such as depression; evidence of Alzheimer's disease and a range of visual impairments from severe visual impairment to total blindness.

A standard eye test at an opticians will be able to provide much information, including about visual acuity and light scattering for both eyes. Screening has often simply involved a visual acuity task where for a pilot, for example, if a normal person can read a chart at 3m, the prospective pilot must be able to read it at 6m (so-called 6/3 vision (reading at 20 feet, what a person with normal vision can read at 20 feet, is 20/20 vision or 6/6 in metres).

Optical devices such as glasses may provide excellent corrections. A judgement must be made in the context of the visual environment of interest whether the person is capable of completing the task and if the tasks and lighting can be modified so that the person can perform at the required level. Of great importance will be safety, and a risk assessment must include whether the person can become a danger to themselves or others.

SCREENING FOR EXPOSURE TO ENVIRONMENTS WITH REDUCED AIR QUALITY

Most health screening criteria for air quality will be in the domain of occupational hygiene and the ability to wear protective clothing. Environmental ergonomics will play a part – for example, heat stress is often a consequence of wearing protective clothing as well as restrictions to vision, warning signals, communication and so on. Environmental ergonomics and air quality are mainly confined to smells and logical considerations would suggest that respiratory disorders, including asthma and allergic reactions, may be the main source of enquiry.

Pre-exposure health screening must correctly negotiate through ethical, equality, privacy and legal obligations. It is crucial, however, to prevent the exposure of people to environments that will damage the health of an individual.

Appendix 3

PHYSICAL MEASURES; OBSERVATION AND SUBJECTIVE SCALES FOR ASSESSING THERMAL ENVIRONMENTS

ISO 28802 (2012) and ISO 10551 (2019) provide the following suggestions for assessing the thermal environment.

Parameters for assessing thermal environments include air temperature; radiant temperature; air velocity; and humidity measured using instruments specified in ISO 7726 (1998) and located at positions representative of the thermal environment experienced by people. When combined with estimates of clothing properties (ISO 9920, 2007) and metabolic heat production (ISO 8996, 2004), the predicted mean vote (PMV) and predicted percentage of dissatisfied (PPD) indices (ISO 7730, 2024), as well as local thermal comfort indices (e.g. draughts), can be calculated (or sweat rate required (SWreq) for hot environments and clothing insulation required (IREQ) for cold environments).

Observation by the investigator includes gaining a general impression of the thermal environment (hot, moderate, cold, etc.), possible causes of local discomfort (e.g. draughts) and of the heating and ventilation system. Occupant behaviour and possible adaptive opportunities (ability to move, adjust clothing or open and close windows, etc.) should be observed. Guidance on risk assessment is provided in ISO 15265 (2004).

Subjective Scales.

Thermal sensation: Please rate on the following scale how YOU feel NOW.

+3, hot; +2, warm; +1, slightly warm; 0, neutral; –1, slightly cool; –2, cool; –3, cold

Extensions are +5, extremely hot; +4, very hot; –4, very cold; –5, extremely cold

Thermal discomfort: In terms of thermal discomfort, please rate on the following scale how uncomfortable YOU feel NOW.

1, not uncomfortable; 2, slightly uncomfortable; 3, uncomfortable; 4, very uncomfortable

Extensions are 5, extremely uncomfortable.

Stickiness: Please rate on the scale how sticky YOU feel now.

1, not sticky; 2, slightly sticky; 3, sticky; 4, very sticky

Extensions are 5, extremely sticky.

Preference scale: In terms of the thermal environment, please rate on the following scale how YOU would like to be NOW.

+1, warmer; 0, no change; -1, cooler

Extended scale: 7, much warmer; 6, warmer; 5, slightly warmer; 4, no change; 3, slightly cooler; 2, cooler; 1, much cooler

Draughtiness: Please rate on the scale how draughty YOU feel now.

1, not draughty; 2, slightly draughty; 3, draughty; 4, very draughty

Extensions are 5, extremely draughty.

Dryness: Please rate on the scale how dry YOU feel NOW.

1, not dry; 2, slightly dry; 3, dry; 4, very dry

Extensions are 5, extremely dry.

Satisfaction: In terms of the thermal environment, please rate on the following scale how YOU feel NOW.

Satisfied/Not satisfied

Acceptability: In terms of the thermal environment, please rate on the following scale how acceptable YOU find the environment to be NOW.

Acceptable/Not acceptable

(Note. YOU and NOW are emphasized to ensure that it is the person's own feeling at that time. Additional requests can replace NOW with IN GENERAL AT WORK as the environmental ergonomist usually has a limited time to conduct an assessment.)

Appendix 4

EXAMPLE OF A ONE-PAGE QUESTIONNAIRE FOR SUBJECTIVE ASSESSMENT OF THE THERMAL ENVIRONMENT

Please rate on the following scales how YOU feel NOW in your thermal environment **Please complete all**

Extremely hot	Very uncomfortable	Very draughty	Very sticky	Very dry
Hot	Uncomfortable	Draughty	Sticky	Dry
Warm	Slightly uncomfortable	Slightly draughty	Slightly sticky	Slightly dry
Slightly warm	Not uncomfortable	Not draughty	Not sticky	Not dry
Neutral				
Slightly cool				
Cool				
Cold				
Extremely cold				

Preference ? Warmer No change Cooler : Acceptable ? Yes No : Satisfied ? Yes No

Please give the reasons for your ratings and any other comments

Please rate on the following scales how YOU feel GENERALLY AT WORK in your thermal environment

Extremely hot	Very uncomfortable	Very draughty	Very sticky	Very dry
Hot	Uncomfortable	Draughty	Sticky	Dry
Warm	Slightly uncomfortable	Slightly draughty	Slightly sticky	Slightly dry
Slightly warm	Not uncomfortable	Not draughty	Not sticky	Not dry
Neutral				
Slightly cool				
Cool				
Cold				
Extremely cold				

Preference ? Warmer No change Cooler : Acceptable ? Yes No : Satisfied ? Yes No

Please give the reasons for your ratings and any other comments

Appendix 5

PHYSICAL MEASURES; OBSERVATION AND SUBJECTIVE SCALES FOR ASSESSING ACOUSTIC ENVIRONMENTS

While recognising the importance of context, ISO 28802 (2012) and ISO 10551 (2019) provide the following suggestions and scales for assessing the acoustic environment.

A-weighted sound pressure level (dB(A)) and equivalent continuous A-weighted sound pressure level dB(A), Leq, should be measured with a sound level meter as specified in IEC 61672-1 and ISO 9612 at a location representative of the acoustic environment of the person. A number of measures for different environmental conditions (background noise only, machines on or off, people present or not), as well as spatial and temporal variations, will provide a 'noise' profile.

Observations by the investigator include a general impression of the acoustic environment as well as of background noise; noise from machines and equipment; printers; footfall; people talking; ventilation fans; and telephones and other relevant sources. Noise frequency and intermittent and impulsive noise, as well as pure tones that can cause annoyance and changes of noise sources throughout the day, should be noted. Whether people are wearing hearing protection can also be useful at the observation stage.

Suggestions for subjective scales are as follows (see also ISO TS 15666, 2021):

Annoyance: Please rate on the following scale how annoying YOU feel the noise in your environment is NOW.

1, not annoying; 2, slightly annoying; 3, annoying; 4, very annoying

Extensions are 5, extremely annoying.

Preference scale: Please rate on the following scale how YOU would like the acoustic environment to be NOW.

1, no change; 2, slightly quieter; 3, quieter; 4, much quieter

Satisfaction: In terms of the acoustic environment, please rate on the following scale how YOU feel NOW.

Satisfied/Not satisfied

Acceptability: In terms of the acoustic environment, please rate on the following scale how acceptable YOU find the environment to be NOW.

Acceptable/Not acceptable

Sources of noise: Please indicate any sources of noise YOU can hear in your environment NOW.

(Note. YOU and NOW are emphasized to ensure it is the person's own feeling at that time. Additional requests can replace NOW with IN GENERAL AT WORK as the environmental ergonomist usually has a limited time to conduct an assessment.)

Appendix 6

EXAMPLE OF A ONE-PAGE SUBJECTIVE QUESTIONNAIRE FOR THE ASSESSMENT OF THE ACOUSTIC ENVIRONMENT

Please rate on the following scales how YOU feel NOW about the sound in YOUR environment **Please complete all**

Extremely annoying	Very uncomfortable	Very noisy	Very interfering	Very distracting
Very annoying	Uncomfortable	Noisy	Interfering	Distracting
Annoying	Slightly uncomfortable	Slightly Noisy	Slightly interfering	Slightly distracting
Slightly annoying	Not uncomfortable	Not Noisy	Not interfering	Not distracting
Not annoying				

Acceptable ? Yes No : Satisfied ? Yes No

Please give the reasons for your ratings and any other comments

Please rate on the following scales how YOU GENERALLY feel AT WORK about the sound in YOUR environment

Extremely annoying	Very uncomfortable	Very noisy	Very interfering	Very distracting
Very annoying	Uncomfortable	Noisy	Interfering	Distracting
Annoying	Slightly uncomfortable	Slightly Noisy	Slightly interfering	Slightly distracting
Slightly annoying	Not uncomfortable	Not Noisy	Not interfering	Not distracting
Not annoying				

Acceptable ? Yes No : Satisfied ? Yes No

Please give the reasons for your ratings and any other comments

Appendix 7

PHYSICAL MEASURES; OBSERVATION AND SUBJECTIVE SCALES FOR ASSESSING VIBRATION ENVIRONMENTS.

While recognising the importance of context, ISO 28802 (2012) and ISO 10551 (2019) provide the following suggestions and scales for assessing the vibration environment.

Weighted rms acceleration level (wrms) and vibration dose value should be measured with accelerometers (ISO 2631, part 1, and ISO 8041) at a location representative of the vibration environment of the person. The frequency-weighting curve used will depend upon the context of the assessment and affects (vehicles, buildings, motion sickness, discomfort, performance, health and hand-transmitted vibration).

Observations by the investigator include a general impression of the vibration environment as well as apparent effects on the ability to carry out tasks, and any distraction and behavioural responses. Motion sickness as well as any avoidance of the environment by individuals and apparent alarm or annoyance should be observed. Adaptive opportunities should be noted such as the ability to change posture as well as move away from the vibration.

Suggestions for subjective scales are as follows:

Discomfort: Please rate on the following scale how uncomfortable YOU feel the vibration in your environment is NOW.

1, not uncomfortable; 2, a little uncomfortable; 3, fairly uncomfortable; 4, uncomfortable; 5, very uncomfortable; 6, extremely uncomfortable.

Satisfaction: In terms of the vibration environment, please rate on the following scale how YOU feel NOW.

Satisfied/Not Satisfied

Acceptability: In terms of the vibration environment, please rate on the following scale how acceptable YOU find the environment to be NOW.

Acceptable/Not acceptable

Sources of vibration: Please indicate any sources of vibration in your environment NOW.

(Note. YOU and NOW are emphasized to ensure it is the person's own feeling at that time. Additional requests can replace NOW with IN GENERAL (adding AT WORK if appropriate) as the environmental ergonomist usually has a limited time to conduct an assessment.)

Appendix 8

EXAMPLE OF A ONE-PAGE QUESTIONNAIRE FOR THE ASSESSMENT OF SUBJECTIVE RESPONSES TO VIBRATION

Please rate on the following scales how YOU feel NOW about the vibration in YOUR environment **Please complete all**

- Extremely uncomfortable
- Very uncomfortable
- Uncomfortable
- Fairly uncomfortable
- A little uncomfortable
- Not uncomfortable

Acceptable ? Yes No :

Please indicate on the diagram where you can feel vibration on your body NOW

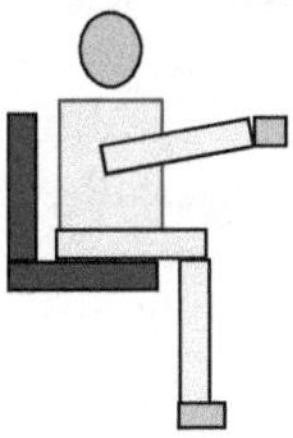

- Very interfering
- Interfering
- Slightly interfering
- Not interfering

Satisfied ? Yes No

- Very distracting
- Distracting
- Slightly distracting
- Not distracting

Please give the reasons for your ratings and any other comments

Appendix 9

PHYSICAL MEASURES; OBSERVATION AND SUBJECTIVE SCALES FOR ASSESSING LIGHTING ENVIRONMENTS

While recognising the importance of context, ISO 28802 (2012) and ISO 10551 (2019) provide the following suggestions and scales for assessing the lighting environment.

Horizontal illuminance (lux) should be measured with a colour-corrected light meter as specified in CIE 69 at a location representative of the lighting and visual environment of the person. A number of measures for different purposes (e.g. task lighting), as well as spatial and temporal variations, will provide a lighting profile. Guidance on the lighting of indoor workplaces can be found in ISO 8995-1.

Observations by the investigator include a general impression of the lighting environment (bright, dim, stimulating, inspirational, etc.) as well as of glare; flicker; reflections; colour; modelling affects and shadows; and decour. Changes of light sources throughout the day should be noted as well as the role of windows and daylight.

Suggestions for subjective scales are as follows:

Discomfort: Please rate on the following scale how much discomfort YOU feel in your visual environment, NOW.

1, no discomfort; 2, slight discomfort; 3, discomfort; 4, much discomfort

Preference scale: Please rate on the following scale how YOU would like your visual environment to be NOW.

1, much darker; 2, darker; 3, slightly darker; 4, no change; 5, slightly lighter; 6, lighter; 7, much lighter

Satisfaction: In terms of the lighting environment, please rate on the following scale how YOU feel NOW.

Satisfied/Not satisfied

Acceptability: In terms of the lighting environment, please rate on the following scale how acceptable YOU find the environment to be NOW.

Acceptable/Not acceptable

Sources of light annoyance: Please indicate any aspects of the visual environment that cause YOU annoyance NOW.

(Note. YOU and NOW are emphasized to ensure it is the person's own feeling at that time. Additional requests can replace NOW with IN GENERAL (AT WORK) as the environmental ergonomist usually has a limited time to conduct an assessment.)

Appendix 10

EXAMPLE OF A ONE-PAGE QUESTIONNAIRE FOR THE SUBJECTIVE ASSESSMENT OF THE VISION AND LIGHTING ENVIRONMENT

Please make a single mark that crosses the line on the scales
to indicate how YOU feel about YOUR vision and lighting environment NOW.

Make a mark on EACH scale

Brightness	Preference	Discomfort and specifics	
Extremely bright	Much darker	Much discomfort	Much flicker
Very bright	Darker	Discomfort	Some flicker
Bright	Slightly darker	Slight discomfort	Slight flicker
Slightly bright	No change	No discomfort	No flicker
Neutral	Lighter	Much glare	Many reflections
Slightly dark	Slightly lighter	Some glare	Some reflections
Dark	Much lighter	Little glare	A few reflections
Very dark		No glare	No reflections
Extremely dark			

In general, Is YOUR vision and lighting environment: Satisfactory? Yes no : Acceptable? Yes no

Please indicate any additional information about how you generally feel about your vision and lighting environment. What are the main good points and bad points? Is the environment distracting? Does it interfere with work? Is it interesting? Is it stimulating?

Appendix 11

PHYSICAL MEASURES; OBSERVATION AND SUBJECTIVE SCALES FOR ASSESSING AIR QUALITY ENVIRONMENTS

While recognising the importance of context, ISO 28802 (2012) and ISO 10551 (2019) provide the following suggestions and scales for assessing the air quality environment.

The level of CO_2 (% or ppm) using a meter specified and approved by a manufacturer provides a method for measuring actual CO_2 which is related to the metabolic activity of people and of combustion processes. It can affect the health, comfort and performance of people, but it is also a surrogate method for assessing effectiveness of ventilation and the possible build-up of other gases, particulates, odours and smells. The level of CO_2 shall be representative of the levels to which people are exposed. In a closed environment, CO_2 distribution will be homogeneous; however, it may vary near windows and doors. In an occupied space with insufficient ventilation, CO_2 levels will build up over the day often causing reports of headaches in the afternoon.

Observations by the investigator include a general impression of the air quality environment, including smells and the causes of smells. Inputs of fresh air and outputs of room air should be noted as well as any building exhaust and intake systems. Inputs near traffic or a car park can be contaminated and inputs near building exhaust areas, air conditioning exhausts or garbage sites and more should be noted as likely to cause contamination. Air circulation patterns, dead spots and the type of ventilation system(s) should be noted. Changes throughout the day should be noted. It should be remembered that people, including the investigator, adapt to smells and that both an impression on first entering a space as well as for extended exposure should be gained.

Suggestions for subjective scales are as follows:

Smelliness: 1, not smelly; 2, slightly smelly; 3, smelly; 4, very smelly
Extension 5, extremely smelly
Satisfaction: In terms of the air quality environment, please rate on the following scale how YOU feel NOW.
Satisfied/Not satisfied
Acceptability: In terms of the air quality environment, please rate on the following scale how acceptable YOU find the environment to be NOW.
Acceptable/Not acceptable
Sources of smells: Please indicate any sources of smell YOU can detect in your environment NOW.

(Note. YOU and NOW are emphasized to ensure it is the person's own feeling at that time. Additional requests can replace NOW with IN GENERAL (AT WORK) as the environmental ergonomist usually has a limited time to conduct an assessment.)

Appendix 12

EXAMPLE OF A ONE-PAGE QUESTIONNAIRE TO ASSESS PERCEIVED AIR QUALITY

Please mark on the scales how YOU perceive the air quality in the environment. **Please mark on all scales**

Overpowering odour	Extremely smelly	Very fresh	Extremely pleasant
Very strong odour	Very smelly	Fresh	Very pleasant
Strong odour	Smelly	Neutral	Pleasant
Moderate odour	Slightly smelly	Slightly stuffy	Slightly pleasant
Slight odour	Not smelly	Stuffy	Not pleasant
No odour			

How would you rate the air quality: Satisfactory: yes no : Unsatisfactory: yes no.

Please provide reasons for your ratings. Is there a main source of smell or odour? Is the Smell reduced after time and is it distracting?

How do you feel NOW?

- Hot
- Warm
- Slightly warm
- Neutral
- Slightly cool
- Cool
- Cold

Appendix 13

EXAMPLE OF A ONE-PAGE QUESTIONNAIRE FOR THE SUBJECTIVE ASSESSMENT OF AN ENVIRONMENT

Mark All scales

Please mark across the line on the scales how you feel YOUR environment is NOW. How do YOU feel?

Thermal

- Extremely hot
- Hot
- Warm
- Slightly warm
- Neutral
- Slightly cool
- Cool
- Cold
- Extremely cold

Noise

- Extremely annoying
- Very annoying
- Annoying
- Slightly annoying
- Not annoying

Vibration

- Extremely uncomfortable
- Very uncomfortable
- Uncomfortable
- Fairly uncomfortable
- A little uncomfortable
- Not uncomfortable

Light

- Extremely bright
- Very bright
- Bright
- Slightly bright
- Neutral
- Slightly dark
- Dark
- Very dark
- Extremely dark

Air quality

- Extremely smelly
- Very smelly
- Smelly
- Slightly smelly
- Not smelly

Circle how YOU find your overall environment? Satisfactory? Yes no : Acceptable : yes no : Quality? high low,

Please indicate why you gave the responses that you did and any other information relevant to YOUR experience in the environment.

References

ACGIH, 2012, TLVs and BEIs, Threshold Limit Values for chemical substances and physical agents & Biological exposure indices, American Conference of Governmental Industrial Hygienists, ISBN 978-1-607260-48-6.

ACGIH, 2024, TLVs and BEIs, Threshold Limit Values for chemical substances and physical agents & Biological exposure indices, American Conference of Governmental Industrial Hygienists, ISBN 978-1-607261-66-7.

ANSI S3.36, 1985, Specification for manikin for simulated in situ airborne acoustic measurements. ANSI, USA.

ANSI/ASHRAE Standard 62.1, 2022, Ventilation and acceptable indoor air quality, ASHRAE, Atlanta, USA.

ANSI/ASHRAE 62.2, 2022, Ventilation in Residential Buildings. ASHRAE, Atlanta, USA.

ANSI/ASHRAE standard 55, 2023, Thermal environmental conditions for human occupancy, Atlanta, USA.

ASHRAE Standard 62 – 89, 1989, Ventilation for acceptable indoor air quality, American Society of Heating, Refrigerating and Air-Conditioning Engineers, Atlanta, USA.

Baker and Standeven, 1997, A behavioural approach to thermal comfort assessment, *International Journal of Solar Energy*, vol 19, pp 21–35.

Bedford, T., 1940, Environmental warmth and its measurement, Medical Research memorandum No 17, HMSO, London.

Bedford, T. and Warner, C. G., 1934, The globe thermometer in studies of heating and ventilation, *Journal of Hygiene*, 34, 458–473.

Bedford, T., 1948, Bedford's basic principles of ventilation and heating, (1st Ed), H K Lewis, London.

Benson, A. J. and Dilnot, S., 1981, Perception of whole-body linear oscillation. Proceedings of the United Kingdom Informal Group on Whole-Body vibration, Heriot-Watt University, Edinburgh. September 9–11, 1981.

Beranek, L., 1971, Noise and vibration control, Cintelez Press, ISBN 978-16825 12203.

Bordas, W., 2002, Learning more from our buildings - or just forgetting less, Review for Building Research and Information journal, Journal of Federal Facilities Council, Technical Report 145, Learning from our buildings, a state of the practice summary of post-occupancy evaluation. Washington National Academy Press, 2001.

Borum, B. F., Schaefer, K. E. and Hastings, B. K., 1954, Effects of exposure to elevated carbon dioxide over a prolonged period, on basal physiological functions and cardiovascular capacity. US Navy Med Res Lab. Report 241.

Boyce, P. R., 2003, Human factors in lighting (2nd Ed), CRC Press, Taylor & Francis Group, London and New York, ISBN 0-7484-0950-5.

Boyce P. R., 2014, Human factors in lighting (3rd Ed), CRC Press, Taylor & Francis Group, London and New York, ISBN 10-978-1439874882.

Braun, T. L. and Parsons, K. C., 1991, Human thermal responses in crowds, In Lovesey E J (Ed) Contemporary Ergonomics, pp 190–195, Taylor & Francis, London.

BS 6841, 1987, Measurement and evaluation of human exposure to whole-body mechanical vibration and repeated shock, BSI, London.

BS 7085, 1989, Guide to safety aspects of experiments in which people are exposed to mechanical vibration and shock, BSI, London.

BS 7000-6, 2005, Design management systems. Managing inclusive design. BSI, London.
BS 7580, 1997, Specification for the verification of sound level meters. www.bsi-global.com
BS EN ISO 5349-1, 2001, Mechanical vibration. Measurement and evaluation of human exposure to hand-transmitted vibration. General requirements, ISO, Geneva.
BS EN ISO 12096, 2001, Mechanical vibration. Declaration and verification of vibration emission values, ISO, Geneva.
BS EN ISO 5349-2, 2002, Mechanical vibration. Measurement and assessment of human exposure to hand-transmitted vibration. Practical guidance for measurement at the workplace. ISO, Geneva.
BS EN 61672-1, 2002, Electroacoustics. Sound level meters. Specifications. www.bsi-global.com
BS EN ISO 8041, 2005, Human response to vibration. Measuring instrumentation. ISO, Geneva.
BS EN 12464-2, 2007, Light and lighting: lighting of workplaces. Outdoor Workplaces, BSI, London, UK.
BS EN 12464-1, 2011, Light and lighting: lighting of workplaces. Indoor Workplaces, BSI, London, UK.
BS EN 12665, 2011, Light and lighting: Basic terms and criteria for specifying lighting requirement, BSI, London, UK.
BS 6164, 2019, Health and safety in tunnelling in the construction industry – code of practice, BSI, London.
Cabanac, M., 1981, Physiological signals for thermal comfort, In Cena K and Clark J A (Eds) Bioengineering, thermal physiology and comfort, Amsterdam, Elsevier.
Canon W B, 1929, Organisation for physiological homeostasis, Physiological Reviews, 1929, 9 (3) pp 399–431.
Carola R, Harley J P and Noback C R, 1992, Human Anatomy, Chapter 16, The Senses, pp 407–442. Mc Graw-Hill, New York +, ISBN 0-07-010527.
CCOHS, 2024, Lighting Ergonomics-Survey and solutions, Canadian Centre for Occupational Health and Safety, ccohs.ca/oshanswers/ergonomcs/lighting/lighting-survey.PDF.
CEN CR 1752:1998, Ventilation for buildings – design criteria for the indoor environment (withdrawn), CEN, Brussels.
Chrenko, F. A., 1974, Bedford's basic principles of ventilation and heating (3rd Ed), H K Lewis, London, ISBN 0-7186-0370-2.
CIBSE, 1992, Lighting Guide, The outdoor environment, Chartered Institution of Building Services Engineers, London, UK.
CIBSE, 1994, Code for interior lighting, Chartered Institution of Building Services Engineers, London, UK.
CIBSE, 1996, Lighting guide LG3: The visual environment for display screen use, Chartered Institution of Building Services Engineers, London, UK.
CIBSE, 2022, Lighting for offices, CIBSE Lighting Guide 7 (LG07), Chartered Institution of Building Services Engineers, London. UK. ISBN 9781914543463.
CIE, 1983, The basis of physical photometry, CIE publication 18.2 CIE, Vienna.
CIE 1997, Low vision lighting needs for the partially sighted, CIE TR 123, CIE (Commission Internationale de l'Eclaraige), Vienna.
CIE, 2024, Methods of characterising illuminance meters and Luminance meters: Performance characteristics and Specifications, International Commission on Illumination (CIE), Vienna.
Clark, R. P. and Edholm, O. G., 1985, Man and his thermal environment, Edward Arnold, London.
Claude Bernard, 1870, Lessons on the phenomenon of life common to animals and vegetables. Second lecture' the three forms of life'. In Langley, L. L. (Ed) Homeostasis: Origins of the concept, 1973. Cited in Ramsay D S and Woods S C, 2014, Clarifying the roles of homeostasis and allostasis in physiological regulation, Psychol Rev. 2014 Apr:121 2); 225-247. DoI 10.1037/a0035942

Coleman, R., 2001, Designing for our future selves. In W. Preiser and D. Ostroff (Eds), Universal design handbook, New York, McGraw-Hill pp 4.1–4.25.

Collins, K. J., 1983, Hypothermia the facts, Oxford University Press, ISBN 0-19-261360-X.

Cooke, H. M., Wyndham, C. H., Strydom, N. B., Maritz, J. S., Bredell, G. A. G., 1961, The effects of heat on the performance of work underground, Journal of the Mine Ventilation, Society of South Africa, October.

Davies, C. N., Davis, P. R. and Tyrer, F. H., 1967, The effects of abnormal physical conditions at work, E & S Livingstone, Edinburgh, London.

Dietz, R. N., Goodrich, E. A. C. and Weiser, R. F., 1986, Detailed description and performance of a passive perfluorocarbon tracer system for building ventilation and air exchange measurement ' Special Technical Publication STP 904, ASTM, Philadelphia, PA, 1986, pp 203–264.

DIN 1946, 1983, *Teil 2:Raumlufttechnic gesundheitsheits-technishe, Anforderungen (VDI-Luftungsregein).* Deutche Institut fur Normuna, Berlin.

Edwards, A. L., 1957, *Techniques of attitude scale construction*, Appleton-century-crofts.INC, New York.

Elgstrand, K. and Petersson, N. F. Eds), 2009, *OSH for development*, Royal Institute of Technology, Sweden, ISBN 978-91-633-4798-6.

Ellis, F. P., Smith, F. E. and Walters, J. D., 1972, Measurement of environmental warmth in SI units, *British Journal of Industrial Medicine*, 29, 361–377.

Elton, E. and Nicolle, C., 2015, Inclusive design and design for special populations. In Wilson J R and Sharples S., *Evaluation of human work* (4th Ed), CRC Press, Taylor & Francis Group, Boca Ratan, New York, Oxford, ISBN 978-1-4665-5961-5, Chapter 11, pp 299–330.

EN 61672-1, 2003, Electroacoustics. Sound level meters. Specifications.

EN 15251, 2007, Indoor environmental input parameters for design and assessment of energy performance of buildings addressing indoor air quality, thermal environment, lighting and acoustics, CEN, Brussels.

EN 16798, 2019, Energy performance of buildings, multi-part standard, (update of EN 15251, 2007), CEN, Brussels.

European Commission, 2000, Directive 2000/14/EC (latest update 2021) of the European Parliament and of the council of 8 May 2000 on noise – equipment for use outdoors, Official Journal of the European Communities.

European Commission, 2002, Directive 2002/44/EC of the European Parliament and of the council of 25 June 2002 on the minimum health and safety requirements regarding the exposure of workers to the risks arising from physical agents (vibration) (sixteenth individual Directive within the meaning of Article 16(1) of Directive 89/391/EEC. Official Journal of the European Communities, L42.

European Commission, 2003/10/EC, Directive of the European Parliament and of the council of 6 February 2003 on the minimum health and safety requirements regarding the exposure of workers to the risks arising from physical agents (noise) (seventeenth individual Directive within the meaning of Article 16(1) of Directive 89/391/EEC. Official Journal of the European Communities, L42.

Fang, L., Clausen, G. and Fanger, P. O., 1998, Impact of temperature and humidity on perception of indoor air quality during immediate and longer whole-body exposures, Indoor Air, 1998, 8, pp 276–284.

European Commission, 2008a, Directive 2008/50/EC of the European Parliament and of the council of 21 May 2008 on ambient air quality and cleaner air for Europe, Official Journal of the European Communities.

European Commission, 2008b, Directive 2008/89/EC of the European Parliament and of the council of 4 September 2008 on the instillation of lighting and light-signalling devices on motor vehicles and their trailers. Official Journal of the European Communities.

Fanger, P. O., 1970, *Thermal Comfort*; Danish Technical Press, Copenhagen, ISBN 87-571-0341-0.

Fanger, P. O., 1988, Introduction of the olf and decipol units to quantify air pollution perceived by humans indoors and outdoors, *Energy and Buildings*, 12, pp 1–6.

Fanger, P. O., Lauridsen, J., Bluyssen, P. and Clausen, G., 1988, Air pollution sources in offices and assembly halls, quantified by the olf unit, *Energy in Buildings*, 12, (1988), pp 7–19.

Fanger, P. O., 1989, The new comfort equation for indoor air quality, IAQ 89; The human equation, Health and comfort, April 17–20, 1989, San Diego.

Feathers, D., D'Souza, C. and Paquet, V., 2015, Anthropometry for Ergonomic design. In ch27, Wilson and Sharples (Eds), 2015, *Evaluation of human work* (4th Ed), CRC Press, Taylor & Francis Group, Boca Ratan, New York, Oxford, ISBN 978-1-4665-5961-5, Chapter 26, pp 705–724.

Fechner, G., 1860, Elements of psychophysics, cited in Guilford J. P., 1954, *Psychometric methods*, (2nd Ed), McGraw-Hill, New York, Toronto, London, ISBN.

Fidell, S., Teffeteller, S. R., Horonjeff, R. D. and Green, G. M., 1979, Predicting annoyance from detectability of low-level sounds. *Journal of the Acoustical Society of America*, 66, 1427–1434.

Field, A. F., 2013, *Discovering statistics using IBM SPSS Statistics* (4th Ed), SAGE, London.

Flemming, D. B. and Griffin, M. J., 1975, A study of the subjective equivalence of noise and whole-body vibration, *Journal of Sound and Vibration*, 42, pp 453–461.

Gagge, A. P., Burton, A. C. and Bazett, H. C., 1941, A practical system of units for the description of the heat exchange of man with his thermal environment, *Science NY*, 94, 428–430.

Gagge, A. P., Stolwijk, J. A. J. and Nishi, Y., 1971, An effective temperature scale based on a single model of human physiological temperature response, *ASHRAE Transactions*, 77, 247–262.

Gallwey, T. J. and O'Sullivan, L. W., 2005, Computer aided ergonomics. In Wilson and Corlett (Eds), *Evaluation of human work*. (3rd Ed) Ch4. pp 83–112, CRC Press, Boco Ratan, FL, ISBN 0-41526757-9.

Gardiner, K., 1995, Noise. In Harrington, J. M. and Gardiner, K., *Occupational Hygiene*, (2nd Ed), Blackwell science, Oxford, ISBN 0-632-03734-2, Chapter 9, pp 123–150.

Gill, F. S., 1995, Ventilation. In Harrington, J. M. and Gardiner, K., *Occupational Hygiene*, (2nd Ed), Blackwell science, Oxford, ISBN 0-632-03734-2, Chapter 22, pp 404–416.

Golden, F. and Tipton, M., 2002, *Essentials of sea survival*, Human kinetics, ISBN 073600215-4

Goldman, R. F., 2006, Thermal manikins, their origins and role, In Fan, J. T. (Ed) Proceedings of the international conference on thermal manikins and modelling, Hong Kong Polytechnic University, ISBN 962-367-534-8.

Goldman, R. F. and Kampman, B., 2007, Handbook of clothing (2nd Ed) *NATO research study group 7 on biomedical aspects of military protective clothing*, Brussels.

Grether, W. F.' Harris, C. S.' Mohr, G. C., Nixon, C. W., et al…, 1971, Effects of combined heat, noise and vibration on human performance and physiological functions, *Aerospace Medicine* 42, pp 1092–1097.

Griffin, M. J., Parsons, K. C., and Whitham, E. M., 1982, Vibration and comfort IV, Application of experimental results, *Ergonomics* 25, 721–739.

Griffin, M. J., 1990, *Handbook of human vibration*; Academic Press, ISBN 0-12-303040-4.

Griffin, M. J., 1995, Vibration. In Harrington, J. M. and Gardiner, K., *Occupational Hygiene* (2nd Ed), Blackwell science, Oxford, ISBN 0-632-03734-2, Chapter 10, pp 151–166.

Guilford, J. P., 1954, *Psychometric methods* (2nd Ed), McGraw-Hill, New York, Toronto, London, ISBN.

Harrington, J. M. and Gardiner, K., 1995, *Occupational Hygiene* (2nd Ed), Blackwell science, Oxford, ISBN 0-632-03734-2.

Haslegrave, C. M., 2015, Auditory environment and noise assessment, In Wilson, J. R. and Sharples, S., *Evaluation of human work* (4th Ed), CRC Press, Taylor & Francis Group, Boca Ratan, New York, Oxford, ISBN 978-1-4665-5961-5, Chapter 26, pp 705–724.

Havenith, G. and Van Middendorp, H., 1990, The relative influence of physical fitness, acclimatization state, anthropometric measures and gender on individual reactions to heat stress, *European Journal of Applied Physiology* 61 5–6): 419–427. DOI:10.1007/BF00236062

Havenith, G., Griggs, K., Qiu, Y., Dorman, L., Kulasekaran, V. and Hodder, S., 2020, Higher comfort temperature preferences for anthropometrically matched Chinese and Japanese versus white-western-middle- European individuals using a personal comfort/cooling system, *Building and Environment*, 183, pp 107–162.

Hignett, S. and Mc Dermott, H., 2015, Methods for collecting and observing participant responses. In Wilson, J. R. and Sharples, S., *Evaluation of human work* (4th Ed), CRC Press, Taylor & Francis Group, Boca Ratan, New York, Oxford, ISBN 978-1-4665-5961-5, Chapter 5, pp 119–138.

Hill, L., Griffith, O. W. and Flack, M., 1916, The measurement of the rate of heat loss by body temperature by convection, radiation and evaporation, *Philosophical Transactions of the Royal Society*, (B), vol 207, pp 183–220.

HMSO, 1945, Environmental Warmth and its measurement, Medical Research Council Memorandum No 17, (Thomas Bedford) HMSO, London.

Hodder, S. G. and Parsons, K. C., 2007, The effects of solar radiation on thermal comfort, *International Journal of Biometeorology*, 51 (3), pp 233–250.

Hopkinson, R. G. and Collins, J. B., 1970, *The ergonomics of lighting, Macdonald technical and scientific*, London, ISBN 356 02680 9.

Howarth, H. V. A. and Griffin, M. J. 1991, The annoyance caused by simultaneous noise and vibration from railways, *Journal of the Acoustical Society of America*, 89 (5), pp 2317–2323.

Howarth, P. A., 2015, Assessment of the visual environment, In Wilson, J. R. and Sharples, S., *Evaluation of human work* (4th Ed), CRC Press, Taylor & Francis Group, Boca Ratan, New York, Oxford, ISBN 978-1-4665-5961-5, Chapter 25, pp 677–704.

HSE, 2005a, Controlling noise at work (2nd Ed), *The control of noise at work regulations, 2005, Health and Safety Executive guidance document*, HSE books, HMSO, UK, ISBN 0-7176-6164-4.

HSE, 2005b, *Hand-arm vibration, The control of vibration at work regulations, 2005, Health and Safety Executive guidance document*, HSE books, HMSO, UK, ISBN 0-7176-6164-4.

HSE, 2005c, *Protect your hearing or lose it, Pack of pocket cards*, HSE Books, ISBN 0-7176-2139-1.

HSE, 2005d, *Whole-body vibration, The control of vibration at work regulations, 2005, Health and Safety Executive guidance document*, HSE books, HMSO, UK, ISBN 0-7176-6126-1.

Huang, Y. and Griffin, M. J., 2012, The effects of sound level and vibration magnitude on the relative discomfort of noise and vibration, *Journal of the Acoustical Society of America*, 131, 6, pp 4558–4569.

Huang, Y. and Griffin, M. J., 2014, The relative discomfort of noise and vibration: effects of stimulus duration, *Ergonomics*, 57, 8, pp 1244–1255.

Humphreys, M. A., 1972, Clothing and thermal comfort of children in secondary schools in summertime, In Proceedings of the CIB Commission W45 Human Requirements) symposium held at the Building Research Station, September 13–16, 1972, Watford, UK.

IEC 61672-1, 2013, Electroacoustics – Sound level meters – Part 1: Specifications, International Electrotechnical commission.

ISO 1996-1, 2003, Acoustics – Description, measurement and assessment of environmental noise: Part 1-Basic quantities and assessment procedures. ISO Geneva.

ISO, 1999, 2013, Acoustics – Estimation of noise-induced hearing loss, (Edition 3, 2013), ISO, Geneva.

ISO 2631-1, 1985, Mechanical vibration and shock. Evaluation of human exposure to whole-body vibration. Part 1: General requirements. ISO Geneva.

ISO 2631-2, 1989 Mechanical vibration and shock. Evaluation of human exposure to whole-body vibration. Part 2: Human exposure to continuous and shock-induced vibrations in buildings (1 to 80 Hz), ISO Geneva.

ISO 2631-1, 1997, Mechanical vibration and shock. Evaluation of human exposure to whole-body vibration. Part 1: General requirements. ISO Geneva.

ISO TR 3352, 1974, Acoustics - Assessment of noise with respect to its effects on the intelligibility of speech, ISO, Geneva.

ISO 3740, 2000, Acoustics – Determination of sound power levels of noise sources – Guidelines for the use of basic standards. ISO, Geneva.

ISO 5349-1, 2001, Mechanical vibration. Measurement and evaluation of hand-transmitted vibration – part 1, General guidelines, ISO Geneva.

ISO 5349-2, 2001, Mechanical vibration. Measurement and evaluation of hand-transmitted vibration – part 2, practical guidance for measurement at the workplace, ISO Geneva.

ISO 6897, 1984, Guidelines for the evaluation of the response of occupants of fixed structures, especially buildings and offshore structures, to low frequency horizontal motion (0.063 to 1 Hz), ISO, Geneva.

ISO 7243, 2017, Ergonomics of the thermal environment – Assessment of heat stress using the WBGT (wet bulb globe temperature) index (3rd Ed) ISO, Geneva.

ISO 7726, 1998, Thermal environments – Instruments and methods for measuring physical quantities, ISO, Geneva.

ISO 7726, 2001, Ergonomics of the thermal environment. Instruments for measuring physical quantities. ISO, Geneva.

ISO 7730, 2005, Ergonomics of the thermal environment – Analytical determination and interpretation of thermal comfort using calculation of the PMV and PPD indices and local thermal comfort criteria, ISO, Geneva.

ISO 7730, 2024, Ergonomics of the thermal environment – Analytical determination and interpretation of thermal comfort using calculation of the PMV and PPD indices and local thermal comfort criteria, ISO Geneva.

ISO 7933, 2004, Ergonomics of the thermal environment – Analytical determination and interpretation of heat stress using calculation of the Predicted Heat Strain, ISO, Geneva.

ISO 7933, 2023, Ergonomics of the thermal environment – Analytical determination and interpretation of heat stress using calculation of the predicted heat strain, (Edition 3, 2023), ISO, Geneva.

ISO 8041, 1990, human response to vibration measuring instrumentation, ISO Geneva.

ISO 8662, 1988 to 1996, *Parts 1 to 14 – Hand-held portable power tools: measurement of vibrations at the handle*, ISO, Geneva.

ISO 8727, 1997, *Mechanical vibration and shock: human exposure, biodynamic coordinate system*, ISO Geneva.

ISO 8995 – 1, 2002, *Lighting of workplaces, Part 1, Indoor*, ISO, Geneva.

ISO 8996, 2004, *Ergonomics of the thermal environment – determination of metabolic rate*, ISO, Geneva.

ISO 9241-210, 2010, Ergonomics of Human-system interaction. Part 210. *Human centred design for interactive systems.* ISO, Geneva.

ISO 9886, 2004, *Evaluation of thermal strain using physiological measurements*, ISO, Geneva.

ISO 9920, 2007, *Estimation of thermal insulation and water vapour resistance of a clothing ensemble*, ISO, Geneva.

ISO 9921, 2003, *Ergonomics – Assessment of speech communication*, ISO Geneva.

ISO 10551, 2019, *Ergonomics of the physical environment* – Subjective judgement scales for assessing physical environments (Edition 2), ISO, Geneva

ISO 11079, 2007, *Ergonomics of the thermal environment – Determination and interpretation of cold stress when using required clothing insulation IREQ and local cooling effects*, ISO, Geneva.

ISO 12894, 2001, *Ergonomics of the thermal environment – Medical supervision of individuals exposed to extreme hot or cold environments*, ISO, Geneva.

ISO 13732-1, 2001, Ergonomics of the thermal environment – Methods for the assessment of human response to contact with surfaces – Part 1: Hot surfaces, ISO, Geneva.

ISO 13732-2, 2001, Ergonomics of the thermal environment – Methods for the assessment of human response to contact with surfaces – Part 2:Human response to contact with surfaces at moderate temperature, ISO, Geneva.

ISO 13732-3, 2005, Ergonomics of the thermal environment – Methods for the assessment of human response to contact with surfaces – Part 3: Cold surfaces, ISO, Geneva.

ISO 14405-1-4, 2024, Ergonomics of the thermal environment – Evaluation of the thermal environment in vehicles. Part 1: Principles; Part 2: Equivalent Temperature; Part 3 assessment using human subjects; Part 4: computer model of the passenger, ISO, Geneva.

ISO 15265, 2004, Ergonomics of the thermal environment – Risk assessment strategy for the prevention of stress or discomfort in thermal working conditions, ISO, Geneva.

ISO TS 15666, 2021, Acoustics – Assessment of noise annoyance by means of social and socio-acoustic surveys, ISO, Geneva.

ISO 15743, 2008, Ergonomics of the thermal environment – Cold work places – Risk assessment and management, ISO, Geneva.

ISO 17772-1, 2017, Energy performance of buildings-indoor environmental quality, Part 1,: Indoor environmental input parameters for the design and assessment of energy performance of building, ISO, Geneva.

ISO/CIE 19476, 2014, Characterization of the performance of illuminance meters and luminance meters, ISO, Geneva.

ISO TR 23454-1, 2024, Human performance in physical environments. Part 1: A performance framework. ISO, Geneva.

ISO 24505, 2016, Ergonomics – Accessible design – Method for creating colour combinations taking account of age-related changes in human colour vision, ISO, Genva.

ISO 28802, 2012, Ergonomics of the physical environment -The assessment of environments by means of an environmental survey involving physical measurements of the environment and subjective responses of people. ISO, Geneva.

ISO 28803, 2012, Ergonomics of the physical environment - Application of international standards to physical environments for people with special requirements, ISO, Geneva.

Jendritzky, G. and de Dear, R. (Eds), 2012, Special issue, Universal Thermal Climate Index (UTCI), *International Journal of Biometeorology*, 56 (3), 419.

Karjalainen, 2012, Thermal comfort and gender: a literature review. *Indoor Air*, 2012, pp 96–109.

Keats, J. A., 1971, *An introduction to quantitative psychology*, John Wiley and son, Sydney, New York, London, Toronto, ISBN 0-471-46240-3.

Kelly, G. A., 1955, *The psychology of personal constructs*, W W Norton and company, New York.

Kelly, L. K., 2011, Thermal comfort on rail journeys, PhD thesis, Department of Human Sciences, Loughborough University, UK.

Kerslake, D. McK 1972, *The stress of hot environments*, Cambridge University Press, UK. ISBN 978-0-521-27936-9.

Keuhner, R. L., 1956, *Humidity effects on the odour problem, ASHRAE Transactions*, 62, ASHRAE, USA, pp 249.

KRISP 2024, Noise cancellation and artificial intelligence, krisp.ai/blog/call-center-background-noise/

Kryter, K. D., 1985, *The effects of noise on man*, (2nd Ed), Academic Press, London, UK.

Kryter, K. D., 1994, *The handbook of hearing and the effects of noise*, Academic Press, San Diego, CA.

Leaman, A., 2003, Post-occupancy evaluation. In Roaf, S. (Ed) *Benchmarking sustainable building performance*, RIBA publications, 2003.

Leithead, C. S. and Lind, A. R., 1964, *Heat stress and heat disorders*, Cassell, London.

Lewis, W. H., 1986, *Underground coal mine lighting handbook*, University of Michigan Library, USA.

Loeb, M., 1986, *Noise and human efficiency*, John Wiley and sons, Chichester, UK.

Mansfield, N. J., 2004, *Human response to vibration*, CRC press, www.crcpress.com, ISBN 0-415-28239-X.

McIntyre, D. A., 1980, *Indoor climate, Applied Science*, London, ISBN 0-85334-868-5.

Monteith, J. L. and Unsworth, M. H., 1990, *Principles of environmental physics* (2nd Ed), Edward Arnold, London, New York, Melbourne, Aukland, ISBN 0-7131-2931-X.

Morioka, M., 1999, Effect of contact location on vibration perception thresholds in the glabrous skin of the human hand. Proceedings of the 34th United Kingdom Group meeting on human response to vibration, held at Ford Motor Company, Dunton, Essex, UK, September 22–24, 1999.

Morioka, M. and Griffin, M. J., 2006, Magnitude-dependence of equivalent comfort contours for fore-and-aft, lateral and vertical whole-body vibration. Journal of Sound and Vibration, 298 (3) 755–772.

Murrell, H., 1973, A history of the ergonomics research society, Ergonomics Research Society publication, booklet to members, 1973, Taylor and Francis, Oxford.

Nelson, C. M., 1997, Hand transmitted vibration assessment: a comparison of results using single axis and triaxial methods. Proceedings of the UK Informal Group on Human Response to Vibration, Institute of Sound and Vibration Research, Southampton, September 17–19, 1997.

Nishi, Y. and Gagge, A. P., 1977, Effective temperature scale useful for hypo- and hyper- baric environments. Aviation space and environmental medicine, 48, pp 97–107.

NKB, 1981, Report No. 41. Indoor Climate, Stockholm.

O'Brian, N. V., Parsons, K. C. and Lamont, D. R., 1996, An assessment of heat strain on workers in tunnels using compressed air compared to those in free air conditions, *In mining and metallurgy, Tunnelling '97 conference*, pp 341–352, London, The Institute of mining and metallurgy.

Olesen, B. W., 1982, *Thermal comfort*, B&K Technical Review, Copenhagen.

Olesen, B. W., 1985, *Local thermal discomfort*, B&K Technical Review No.1, Copenhagen.

Olesen, B. W. and Dukes-Dubos, F., 1988, International standards for assessing the effects of clothing on heat tolerance and comfort, In Mansdorf, S. Z., Sager, R. and Nielson, A. R. (Eds) *Performance of protective clothing*, pp 17–30, ASTM, Philadelphia.

Ollerhead, J. B., 1973, Noise: How can nuisance be controlled, *Applied Ergonomics*, 4(3), 130–138.

Pandolf, K. B., Sawka, N. M. and Gonzalez, R. R., 1988, *Human performance physiology and environmental medicine at terrestrial extremes*, Brown and Benchmark, USA, ISBN 0-697-14823-8.

Parker, R. D. and Parsons, K. C., 1990, Computer based system for the estimation of clothing insulation and metabolic heat production, in E. J. Lovesey (Ed.), *Contemporary Ergonomics 1990*, London: Taylor & Francis, pp. 473–478.

Parsons, K. C., 2018, ISO standards on physical environments for worker performance and productivity. Ind Health, 56(2):93–95, doi: 10.2486/indhealth.56-93.

Parsons, K. C. and Griffin, M. J., 1983, Methods for predicting passenger vibration discomfort, SAE Technical Paper, 831029, Passenger car meeting, Dearborn, Michigan, June 6–9, 1983.

Parsons, K. C. and Griffin, M. J., 1988, Whole-body vibration perception thresholds, *Journal of Sound and Vibration*, 121 (2), pp 237–258.

Parsons, K. C. and Bishop, D., 1991, A data base model of human responses to thermal environments, in E. J. Lovesey (Ed.), *Contemporary Ergonomics*, London: Taylor & Francis, pp. 444–449.
Parsons, K. C., 1992, The thermal audit, In Lovesey, E. J., (Ed) *Contemporary Ergonomics*, pp 85–90, Taylor & Francis, London.
Parsons, K. C., 1995, Ergonomics. In Harrington, J. M. and Gardiner, K., *Occupational Hygiene*, 2nd Ed), Blackwell science, Oxford, ISBN 0-632-03734-2, Chapter 15, pp 261–275.
Parsons, K. C., 2000, Environmental ergonomics: A review of principles, methods and models, *Applied Ergonomics*, 31, 6, pp 581–594.
Parsons, K. C. and Mahudin, N. D. M., 2004, Development of a crowd stress index (CSI) for use in risk assessment, In McCabe, P. T. (Ed), *Contemporary Ergonomics*, pp 410–440, Boca Raton, FL. CRC Press.
Parsons, K. C. 2014, *Human thermal environments* (3rd Ed) CRC Press, Taylor & Francis Group, Boc Ratan, Oxford, New York ISBN 978-1-4665-9599-6.
Parsons, K. C., 2015a, Ergonomics assessment of thermal environments, In Wilson, J. R. and Sharples, S., *Evaluation of human work* (4th Ed), CRC Press, Taylor & Francis Group, Boca Ratan, New York, Oxford, ISBN 978-1-4665-5961-5, Chapter 24, pp 655–676.
Parsons, K. C., 2015b, The environmental ergonomics survey, In Wilson J R and Sharples S., *Evaluation of human work* (4th Ed), CRC Press, Taylor & Francis Group, Boca Ratan, New York, Oxford, ISBN 978-1-4665-5961-5, Chapter 23, pp 641–654.
Parsons, K.C., 2019, *Human heat stress*, CRC Press, Taylor & Francis Group, Boca Ratan, London, New York, ISBN 978-0-367-00233-6.
Parsons, K. C., 2020, *Human thermal comfort*, CRC Press, Boca Raton, ISBN 978-0-367-26193-1
Parsons, K.C., 2021, *Human cold stress*, CRC Press, Taylor & Francis Group, Boca Ratan, London, New York, ISBN 978-0-367-55199-5.
Pheasant, S. and Haselgrave, C. M., 2006, *Bodyspace: Anthropometry, Ergonomics and the design of work*, (3rd Ed), CRC Press, Taylor & Francis, Boca Raton FL, USA.
Piasecki, M., Kostyrko, K. Fedorczak-Cisak, M. and Nowak, K., 2020, Air enthalpy as an IAQ indicator in hot and humid environment – experimental evaluation, *Energies*, 13, 6, 1481.
Raleigh, V., 2024, What is happening to life expectancy in England, Kingsfund.org.UK
Ramsay, D. S. and Woods, S. C., 2014, Clarifying the roles of homeostasis and allostasis in physiological regulation, *Psychol Rev*. Apr:121 2); 225–247. DoI 10.1037/a0035942
Rappaport, M. B. and Szocik, K. (Eds), 1986, Lewis W H. Underground coal mine lighting handbook, Part 2: Application. Bureau of Mines information circular IC 9073, Pittsburgh, USA.
Reason, J., 1974, *Man in motion*, Littlehampton book services Ltd, ISBN 978-0297766889
Reid, D. B., 1844, *Illustrations of the theory and practice of ventilation*, Longman, Brown, Green and Longmans, London.
Richards, L. G., Jacobson, I. D., Barber, R. W. and Pepler, R. C., 1978, Ride quality evaluation in ground based vehicles: Passenger comfort models for buses and trains, *Ergonomics*, 21, 463–472.
Schultz, J., 1978, Synthesis of social surveys on noise annoyance, *Journal of the Acoustical Society of America*, 64, 377–405.
Shadbolt, N. R. and Smart, P. R., 2015, Knowledge elicitation: methods, tools and techniques. In Wilson and Sharples Eds, *Evaluation of human work*. ch 7, pp 163–200.
Seidel, H. and Heide, R., 1986, Long term effects of whole-body vibration: a critical survey of the literature, *International Archives of Occupational and Environmental Health*, 58, pp 1–26.
Sharples, S., 2015, Assessment and design of the physical workplace, Introduction to Part IV, In Wilson, J. R. and Sharples, S., *Evaluation of human work* (4th Ed), pp 139–140, CRC Press, Taylor & Francis Group, Boca Ratan, New York, Oxford, ISBN 978-1-4665-5961-5.

Sharples, S. and Cobb, S., 2015, Methods for collecting and observing participant responses. In Wilson, J. R. and Sharples, S., *Evaluation of human work* (4th Ed), CRC Press, Taylor & Francis Group, Boca Ratan, New York, Oxford, ISBN 978-1-4665-5961-5, Chapter 4, pp 61–82.

Siegel, S., 1956, *Non-parametric statistics for the behavioural sciences*, McGraw-Hill Kogakusha Ltd, Tokyo.

Sinclair, M. A., 2005, Participative assessment In Wilson and Corlett (Eds), *Evaluation of human work* (3rd Ed). Ch4. pp 83–112, CRC Press, Boco Ratan, FL, ISBN 0-41526757-9.

Smith, T. A. and Parsons, K. C., 1987, The design, development and evaluation of a climatic ergonomics knowledge-based system. In Megaw, E. D. (Ed), *Contemporary Ergonomics*, pp 257–262, London: Taylor & Francis.

Smith, N. A., 1995, Light and lighting. In Harrington, J. M. and Gardiner, K., *Occupational Hygiene*, (2nd Ed), Blackwell Science, Oxford, ISBN 0-632-03734-2, Chapter 11, pp 167–190.

Snow, C. E., 1927, Research on industrial illumination Tech. *Eng. News*, 8, pp 257–282.

Stevens, S. S., 1951, (Ed), *Handbook of experimental psychology*, Wiley, New York.

Stevens, S. S., 1957, On the psychophysical law, *Pschol Rev*, 64, 153–181.

Stevens, S. S., 1975, *Psychophysics*, John Wiley and sons, ISBN 9781429021203.

Stolwijk, J. A. J. and Hardy, J. D., 1977, Control of body temperature, in Handbook of Physiology, section 9: reaction to environmental agents, *Bethesda, Maryland: American Physiological Society*, pp. 45–68.

Sulaman, F. G., 1974, Meteorological front movements and human weather sensitivity, *Karger Gazette*, vol 30, pp 1–6.

Thompson, A. M. S., House, R., Krajnak, T. and Eger, T., 2010, Vibration white foot – a case report, *Occupational Medicine*, vol 60, (7), pp 572–574 Thurstone 1927 – see Edwards, 1957).

Toftum, J. and Fanger, P. O., 1999, Air humidity requirements for human comfort, *ASHRAE Transactions*, 105, 81–86.

University of Strathclyde, 2010, *Post Occupancy Evaluation Questionnaire*. https://www.esru.strath.ac.uk>EandE>survey2

Vincent, J. H. and Brosseau, L. M., 1995, In Harrington, J. M. and Gardiner, K., (Eds) *Occupational Hygiene* (2nd Ed) Blackwell science, Oxford, ISBN 0-632-03734-2.

Wadsworth, P. M. and Parsons, K. C., 1989, The design, development, evaluation and implementation of an expert system into an organisation. In Proceedings of the 3rd International conference on Human-computer interaction, Elsevier, Amsterdam.

Webster, J. C., (1979), Effects of noise on speech. In. Handbook of Noise Control. 2nd Edition, C. M. Harris. Ed. New York, McGraw-Hill.

Wilson, J. R. and Sharples, S., 2015a, Methods in the understanding of human factors. In Wilson, J. R. and Sharples, S., *Evaluation of human work* (4th Ed), CRC Press, Taylor & Francis Group, Boca Ratan, New York, Oxford, ISBN 978-1-4665-5961-5, Chapter 1, pp 1–36.

Wilson, J. R. and Sharples, S., 2015b, Evaluation of human work (4th Ed) CRC Press, Taylor & Francis Group, Boca Ratan, New York, Oxford, ISBN 978-1-4665-5961-5.

Wyon, D. P. and Holmberg, I., 1972, Systematic observation of classroom behaviour during moderate heat stress, In Thermal comfort in moderate heat stress, Proceedings of the CIB Commission W45 symposium held at the Building Research Station, Watford, UK.

Yaglou, C. P. and Minard, D., 1957, Control of heat casualties at military training centers, *American Medical Association Archives of Industrial Health*, 16, 302–316 and 405.

Yaglou, C. P., Riley, C. P. and Coggins, D. I., 1936, Ventilation requirements, Pt 1, *ASHVE Transactions*, pp 133–162.

Yegeneh, B., Haghihat, F., Gunnarsen, L. B., Afshari, K. and Knudsen Nellemose, H. N., 2006, Evaluation of building materials individually and in combination using odour threshold, *Indoor and Built Environment* 15(6), 583–593.

Youle, A., 1995, The thermal environment. In Harrington J M and Gardiner K., *Occupational Hygiene* (2nd Ed), Blackwell science, Oxford, ISBN 0-632-03734-2, Chapter 12, pp 191–226.

Index

Pages in *italics* refer to figures and pages in **bold** refer to tables.

J

K

L

M

Q

R

S

T

U

V

W

X

Y

Z